Biotechnology and Cloning

ISSUES

First published by Independence

The Studio, High Green

Great Shelford

Cambridge CB22 5EG

England

© Independence 2011

Copyright

This book is sold subject to the condition that it shall not, by way of trade or otherwise, be lent, resold, hired out or otherwise circulated in any form of binding or cover other than that in which it is published without the publisher's prior consent.

Photocopy licence

The material in this book is protected by copyright. However, the purchaser is free to make multiple copies of particular articles for instructional purposes for immediate use within the purchasing institution. Making copies of the entire book is not permitted.

British Library Cataloguing in Publication Data

Biotechnology and cloning. -- (Issues ; v. 211)

1. Cloning--Moral and ethical aspects.

I. Series II. Firth, Lisa.

176-dc22

ISBN-13: 978 1 86168 587 2

Printed in Great Britain

MWL Print Group Ltd

CONTENTS

Chapter 1 Biotechnology

What is biotechnology?	1
Emerging biotechnologies	2
Europeans and biotechnology in 2010: winds of change?	5

Chapter 2 Animal Cloning

Animal cloning	7
Timeline of domestic species cloned	9
Q&A on cloning of animals for food	10
Cloned cows	12
'If anything, this milk will be better quality',	13
Food watchdog admits tracking cloned cows 'impossible'	15
Cloned animals and their offspring	17
Cloned British meat is 'safe'	18
More evidence required on cloning	19
In vitro meat	20
Why I'd happily eat lab-grown meat	21

Chapter 3 Human Cloning

Therapeutic cloning (somatic cell nuclear transfer)	22
Reproductive cloning	24
What are stem cells?	28
Stem cell research: hope or hype?	29
First 'saviour sibling' stem cell transplant performed in UK	32
Aborted fetal tissue used in stem cell trial – no thank you	33
New ethical challenges in stem cell research	34
Look, no embryos! The future of ethical stem cells	35
'I travelled in hope to stem cell clinic'	38
Key Facts	40
Glossary	41
Index	42
Acknowledgements	43
Assignments	44

OTHER TITLES IN THE ISSUES SERIES

For more on these titles, visit: www.independence.co.uk

Vegetarian and Vegan Diets ISBN 978 1 86168 406 6
Sustainability and Environment ISBN 978 1 86168 419 6
The Terrorism Problem ISBN 978 1 86168 420 2
Religious Beliefs ISBN 978 1 86168 421 9
A Classless Society? ISBN 978 1 86168 422 6
Migration and Population ISBN 978 1 86168 423 3
Climate Change ISBN 978 1 86168 424 0
Euthanasia and the Right to Die ISBN 978 1 86168 439 4
Sexual Orientation and Society ISBN 978 1 86168 440 0
The Gender Gap ISBN 978 1 86168 441 7
Domestic Abuse ISBN 978 1 86168 442 4
Travel and Tourism ISBN 978 1 86168 443 1
The Problem of Globalisation ISBN 978 1 86168 444 8
The Internet Revolution ISBN 978 1 86168 451 6
An Ageing Population ISBN 978 1 86168 452 3
Poverty and Exclusion ISBN 978 1 86168 453 0
Waste Issues ISBN 978 1 86168 454 7
Staying Fit ISBN 978 1 86168 455 4
Drugs in the UK ISBN 978 1 86168 456 1
The AIDS Crisis ISBN 978 1 86168 468 4
Bullying Issues ISBN 978 1 86168 469 1
Marriage and Cohabitation ISBN 978 1 86168 470 7
Our Human Rights ISBN 978 1 86168 471 4
Privacy and Surveillance ISBN 978 1 86168 472 1
The Animal Rights Debate ISBN 978 1 86168 473 8
Body Image and Self-Esteem ISBN 978 1 86168 484 4
Abortion – Rights and Ethics ISBN 978 1 86168 485 1
Racial and Ethnic Discrimination ISBN 978 1 86168 486 8
Sexual Health ISBN 978 1 86168 487 5
Selling Sex ISBN 978 1 86168 488 2
Citizenship and Participation ISBN 978 1 86168 489 9
Health Issues for Young People ISBN 978 1 86168 500 1
Crime in the UK ISBN 978 1 86168 501 8
Reproductive Ethics ISBN 978 1 86168 502 5
Tackling Child Abuse ISBN 978 1 86168 503 2
Money and Finances ISBN 978 1 86168 504 9
The Housing Issue ISBN 978 1 86168 505 6
Teenage Conceptions ISBN 978 1 86168 523 0
Work and Employment ISBN 978 1 86168 524 7
Understanding Eating Disorders ISBN 978 1 86168 525 4
Student Matters ISBN 978 1 86168 526 1
Cannabis Use ISBN 978 1 86168 527 8
Health and the State ISBN 978 1 86168 528 5
Tobacco and Health ISBN 978 1 86168 539 1
The Homeless Population ISBN 978 1 86168 540 7
Coping with Depression ISBN 978 1 86168 541 4
The Changing Family ISBN 978 1 86168 542 1
Bereavement and Grief ISBN 978 1 86168 543 8
Endangered Species ISBN 978 1 86168 544 5
Responsible Drinking ISBN 978 1 86168 555 1
Alternative Medicine ISBN 978 1 86168 560 5
Censorship Issues ISBN 978 1 86168 558 2
Living with Disability ISBN 978 1 86168 557 5
Sport and Society ISBN 978 1 86168 559 9
Self-Harming and Suicide ISBN 978 1 86168 556 8
Sustainable Transport ISBN 978 1 86168 572 8
Mental Wellbeing ISBN 978 1 86168 573 5
Child Exploitation ISBN 978 1 86168 574 2
The Gambling Problem ISBN 978 1 86168 575 9
The Energy Crisis ISBN 978 1 86168 576 6
Nutrition and Diet ISBN 978 1 86168 577 3
Coping with Stress ISBN 978 1 86168 582 7
Consumerism and Ethics ISBN 978 1 86168 583 4
Genetic Modification ISBN 978 1 86168 584 1
Education and Society ISBN 978 1 86168 585 8
The Media ISBN 978 1 86168 586 5
Biotechnology and Cloning ISBN 978 1 86168 587 2

A note on critical evaluation

Because the information reprinted here is from a number of different sources, readers should bear in mind the origin of the text and whether the source is likely to have a particular bias when presenting information (just as they would if undertaking their own research). It is hoped that, as you read about the many aspects of the issues explored in this book, you will critically evaluate the information presented. It is important that you decide whether you are being presented with facts or opinions. Does the writer give a biased or an unbiased report? If an opinion is being expressed, do you agree with the writer?

Biotechnology and Cloning offers a useful starting point for those who need convenient access to information about the many issues involved. However, it is only a starting point. Following each article is a URL to the relevant organisation's website, which you may wish to visit for further information.

BIOTECHNOLOGY

Chapter 1

What is biotechnology?

Information from The Association of the British Pharmaceutical Industry (ABPI).

Biotechnology can be described in many different ways, but a straightforward definition is: 'Biotechnology is the use of biological organisms or enzymes in the synthesis, breakdown or transformation of materials in the service of people.'

The polymerase chain reaction has enabled scientists to make large quantities of DNA from tiny samples. This in turn has made much of the most recent and exciting DNA technology possible. The ability to sequence DNA, identifying genes and what they do, the Human Genome Project leading on to the 1000 Genomes Project, and DNA fingerprinting – all depend for their success on this ingenious technique.

One area of biotechnology which nearly everyone has heard of is genetic engineering. We now have genetically-modified organisms ranging from bacteria to cows and sheep which produce life-saving medicines including vaccines, insulin and blood-clotting factors. Genetically-modified bacteria can even make the lung surfactant needed to save the life of a premature baby. Hundreds of thousands of people benefit from the chemicals these very special organisms produce.

Gene therapy is another area of medical biotechnology which is still in the early stages of development. The hope is that gene technology will help scientists develop ways to correct mistakes in the DNA code which lead to other genetic diseases in the way they have for SCID (severe combined immunodeficiency).

Medicine also benefits from many sensitive tests which indicate the presence or absence of substances in body fluids. Biotechnological advances in the use of immobilised enzymes and monoclonal antibodies mean these tests have become increasingly rapid and accurate in recent years. A common example is the pregnancy test. This used to take weeks – now it can be done at home on the first day of a missed period and the results are ready in minutes!

More controversial developments in biotechnology involve research into the use of stem cells. Embryonic stem cells are taken from the hollow ball of cells which make up the early human embryo. They have the potential to grow and develop into new tissues or organs to replace others which are worn out or diseased. The use of human embryos means that it is, for some people, very controversial.

Most of the developments in biotechnology have taken place in a very short space of time, beginning with the revelation of the structure of the DNA molecule nearly 60 years ago. The biotechnology timeline shows just how rapidly DNA technology has developed.

⇨ The above information is reprinted with kind permission from the Association of the British Pharmaceutical Industry (ABPI). Visit www.abpischools.org.uk for more information.

© *ABPI*

ISSUES: BIOTECHNOLOGY AND CLONING

chapter one: biotechnology

ASSOCIATION OF THE BRITISH PHARMACEUTICAL INDUSTRY

Emerging biotechnologies

Information from the Nuffield Council on Bioethics.

To describe a technology as 'emerging' might simply be to imply that it is relatively new and not, as yet, accepted into routine use. However, the term 'emerging technology' has acquired a more complex meaning that, despite its increasingly widespread use, is difficult to define precisely.

'Emerging technologies' have been described as ones that:

⇨ arise from new knowledge, or the innovative application of existing knowledge;

⇨ lead to the rapid development of new capabilities;

⇨ are projected to have significant systemic and long-lasting economic, social and political impacts;

⇨ create new opportunities for and challenges to addressing global issues; and

⇨ have the potential to disrupt or create entire industries.

While this is only one possible description and different perspectives will draw attention to different features, it shows how most references to emerging technologies imply something more than merely the incremental development of a new invention, technique or process from proof of concept through testing and validation to acceptance into routine use.

Emerging biotechnologies – the subject of our consultation – are those emerging technologies with a biological basis or use. Without settling on a definition of emerging biotechnologies for the time being we can nevertheless suggest certain features that may characterise them (although not necessarily all of them) and by which they might be recognised.

The developmental pathway followed by these technologies may vary considerably. Some emerging biotechnologies that were conceived decades ago have taken a long time or have not (yet) achieved their original promise (e.g. xenotransplantation) while others have been implemented relatively quickly (e.g. preimplantation genetic diagnosis); some are in early stages of use with uncertain prospects (e.g. genomic medicine) while others are at early stages of research (e.g. synthetic biology) with an uncertain range of uses.

> **Some emerging biotechnologies that were conceived decades ago have taken a long time or have not (yet) achieved their original promise**

Emerging biotechnologies may range from techniques (with, initially, a single, specific and limited use) to programmes (which encompass many techniques). Techniques may emerge by finding a variety of, often highly diverse, uses (such as polymerase chain reaction, which underpins genetic testing from healthcare to law enforcement and genealogy, or in vitro fertilisation, which underpins preimplantation genetic testing for diverse a range of purposes). Programmes, on the other hand, may draw on many different techniques (for example, human enhancement technologies, which draw on distinct innovations in pharmacology, surgery, computing, engineering, and cellular medicine among other things). This may present problems for ethical, policy, funding and regulatory frameworks, in that they may be applied to a technique in one of its many uses but not others, or to a programme but not to all of the techniques that underpin it.

Typically, there are significant uncertainties about the risks associated with an emerging biotechnology and considerable dispute about these. This may be due to its novelty, making it hard to draw inferences about risk by analogy to familiar technologies, or simply because the risks have not yet been well researched. This can lead to

exaggerated claims both for and against the technology, and criticism of its proponents or opponents. Whether a technology is considered to be 'emerging' or not may be related to whether it is demonstrated to raise particular issues, whether it might raise such issues or whether (and by whom) it is claimed to raise those issues.

There are significant uncertainties about the risks associated with an emerging biotechnology and considerable dispute about these

Often, emerging biotechnologies have the capacity to generate significant controversy (for example genetically modified crops or synthetic biology). This may be because of uncertainty or disagreement about the nature and level of associated risks. It also may be because they disturb familiar distinctions that are associated with established values, such as the distinction between the natural and the artificial (as when cloned animals are introduced into the food chain) or between human and non-human (as with the creation of human-animal 'admixed' embryos). Controversies can be particularly heightened where there is a possibility that the technology could be used for anti-social or criminal purposes.

Emerging biotechnologies may present important decisions for policy makers because they entail a commitment of resources or foreclose the development of an alternative approach. Alternatively, bans or over-restrictive regulation can lead to the delay or loss of valuable benefits. In some cases, the significance of a policy decision, or even that the opportunity for making a decision was present at all, may only be recognised after the fact, by which time it may have become difficult to reverse.

Emerging biotechnologies often have an inter-disciplinary character, drawing on knowledge originating in a variety of more established fields. Novelty (of approach and perspective), convergence (between fields and practices), and divergence (from an established 'parent' field) may be characteristic features. Emerging technologies tend to be associated with the development of new concepts and give rise to new kinds of problem or 'unlock' new perspectives. They tend to be fertile sources of novel technical terms (for example, the concept of a 'biobrick' in synthetic biology).

Ethical issues

The emergence of new biotechnologies gives rise to questions about whether, in what conditions, and subject to what controls they should be put to use: in other words, questions of ethics and public policy.

Ethics is a branch of philosophy that concerns what we ought or ought not to do; it involves the study of values and moral reasoning, and their application to human conduct. Bioethics is a branch of applied ethics that deals with moral problems arising from the life sciences.

Below, we describe a number of ethical issues that we think may be relevant to emerging biotechnologies. We do not wish to suggest that this is an exhaustive list and readers may reject, elaborate or add to these.

Human intervention in nature

Concerns have been raised about the capacity of certain biotechnologies, such as genetic engineering and synthetic biology, to change the relationship between human beings and the natural world in morally undesirable ways. There are concerns that these biotechnologies inappropriately elevate human agency (so-called 'playing God'), instrumentalise other living beings, create a capacity for intervention in the natural world that is undesirable in itself, and disrupt distinctions to which people attach moral significance (such as that between humans and non-humans).

Harms to health

Ethical issues arise when the use of technologies is believed, by at least some, to be potentially harmful to the health of individuals or populations, or to have

a capacity for harm as well as benefit. It has been suggested that nanotechnology, for example, has this capacity. It is important to consider the significance of such beliefs and the respect that is due to them, as well as what can be known about the balance between potential harm and potential benefits arising from the technologies in question, and how these depend on context and perspective.

There are concerns that some emerging biotechnologies may widen the gap between the rich and the poor

Environmental harm

In addition to potentially detrimental effects on human wellbeing, the effect a certain technology (e.g. genetically modified [GM] crops) might have on the environment, natural ecologies and other living things (for example, habitat destruction or pollution) may also give rise to significant ethical concerns.

Human nature

Technologies that are perceived to challenge common understanding of what it is to be human often give rise to controversy. For example, xenotransplantation or human enhancement technologies that might, in the future, significantly alter human cognitive or physical capacities, might eventually call into question how we identify a given person as human. Developments like these can challenge value systems that attach to concepts such as human integrity, human nature and human dignity.

Social and intergenerational justice

There are concerns that some emerging biotechnologies may widen the gap between the rich and the poor, at both a national level (e.g. the emergence of a genetic underclass of people who are excluded from certain forms of social participation because of their genetic inheritance) and an international level (e.g. developing economies being excluded from benefit sharing or lacking a knowledge base, facilities and/or funding for licensed products or components).

Some emerging biotechnologies, such as GM crops or synthetic biology, have global implications: they are likely to involve multinational companies and the transfer of money, products and people between countries. However, this may become unavoidable if, for example, advanced agricultural biotechnologies hold the only feasible long-term approach to maintaining an adequate food supply in some parts of the world.

The consequences of implementing (or not implementing) new technologies may have an impact on future generations (e.g. spent nuclear fuel storage). Where this is the case we need to consider the obligations that the present generation owes to those that follow it and the risks and benefits of new developments for future generations.

April 2011

⇨ The above information is an extract from *Emerging biotechnologies: consultation paper* and is reprinted with kind permission from the Nuffield Council on Bioethics. Visit www.nuffieldbioethics.org/biotechnologies for more information.

© *Nuffield Council on Bioethics*

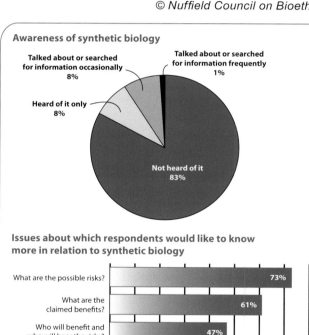

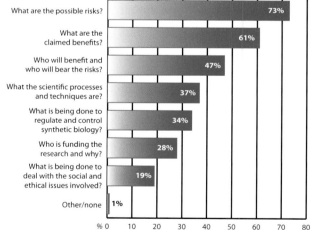

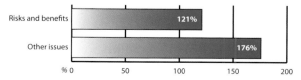

Note: Percentages sum to 300 because respondents have chosen three pieces of information. Only those respondents who chose three pieces of information are included in this graph: those who responded 'don't know' or who mentioned only one or two types of information are excluded.

Source: Europeans and Biotechnology in 2010: Winds of change? http://ec.europa.eu, © European Union, 1995-2011

Europeans and biotechnology in 2010: winds of change?

Overview of key findings.

The latest Eurobarometer survey on the Life Sciences and Biotechnology, based on representative samples from 32 European countries and conducted in February 2010, points to a new era in the relations between science and society. While entrenched views about GM food are still evident, the crisis of confidence in technology and regulation that characterised the 1990s – a result of BSE, contaminated blood and other perceived regulatory failures – is no longer the dominant perspective. In 2010 we see a greater focus on technologies themselves: are they safe? Are they useful? And are there 'technolite' alternatives with more acceptable ethical-moral implications? Europeans are also increasingly concerned about energy and sustainability. There is no rejection of the impetus towards innovation: Europeans are in favour of appropriate regulation to balance the market, and wish to be involved in decisions about new technologies when social values are at stake.

Overall, Europeans consider synthetic biology a sensitive technology that demands precaution and special regulations

Technological optimism

A majority of Europeans are optimistic about biotechnology (53 per cent optimistic; 20 per cent say 'don't know'). In comparison, they are more optimistic about brain and cognitive enhancement (59; 20), computers and information technology (77; 6), wind energy (84; 6) and solar energy (87; 4), but are less optimistic about space exploration (47; 12), nanotechnology (41; 40) and nuclear energy (39; 13). Time series data on an index of optimism show that energy technologies – wind energy, solar energy and nuclear power – are on an upward trend – what we call the 'Copenhagen effect'. While both biotechnology and nanotechnology had seen increasing optimism since 1999 and 2002 respectively, in 2010 both show a similar decline – with support holding constant but increases in the percentages of people saying they 'make things worse'. With the exception of Austria, the index for biotechnology is positive in all countries in 2010, indicating more optimists than pessimists – Germany joining Austria in being the least optimistic about biotechnology. But in only three countries (Finland, Greece and Cyprus) do we see an increase in the index from 2005 to 2010.

Synthetic biology

Following a description of synthetic biology, respondents in the survey were asked: 'Suppose there was a referendum about synthetic biology and you had to make up your mind whether to vote for or against. Among the following, what would be the most important issue on which you would like to know more?' Our respondents were asked to select three from the list of seven issues of interest. 73 per cent selected 'possible risks'; 61 per cent 'claimed benefits' and 47 per cent 'who will benefit and who will bear the risks'. Information about social and ethical issues was the least frequent choice at 19 per cent. Asked about their views on whether, and under what conditions, synthetic biology should be approved, of those respondents who expressed a view 17 per cent said that they do not approve under any circumstances; 21 per cent do not approve except under very special circumstances; 36 per cent approve as long as synthetic biology is regulated by strict laws; three per cent approve without any special laws.

Overall, Europeans consider synthetic biology a sensitive technology that demands precaution and special regulations, but an outright ban would not find overwhelming support.

GM food

GM food is still the Achilles' heel of biotechnology. The wider picture is of declining support across many of the EU Member States – on average opponents outnumber supporters by three to one, and in no country is there a majority of supporters. What is driving the continued opposition to GM food? Public concerns about safety are paramount, followed by the perceived absence of benefits and worry – GM food is seen as unnatural and makes many Europeans 'uneasy'. Across the period 1996-2010, we see, albeit with fluctuations, a downward trend in the percentage of supporters. Denmark and the UK, at the higher end of the distribution of support, are exceptions, as is Austria, at the lower end. Those among the 'old' EU countries with a ban on GM crops in place consistently show low values of support, with Italy joining the group. In contrast, Member States where GM crops are grown tend to show among the highest values, suggesting a link between private attitudes and public policies.

Animal cloning for food products

Cloning animals for food products is even less popular than GM food, with 18 per cent of Europeans in support. In only two countries – Spain and the Czech Republic – does animal cloning attract the support of three in ten. This contrasts with 14 countries in which support for GM food is above 30 per cent. Is this an indication of broader public anxieties about biotechnology and food? The idea of the 'natural superiority of the natural' captures many of the trends in European food production, such as enthusiasm for organic food, local food, and worries about food miles. And if 'unnaturalness' is one of the problems associated with GM food, it appears to be an even greater concern in the case of animal cloning and food products.

Regenerative medicine

Developments in regenerative medicine attract considerable support across Europe. 68 per cent of respondents approve of stem cell research and 63 per cent approve of embryonic stem cell research. Levels of approval for gene therapy are similar, at 64 per cent. Xenotransplantation – an application long subject to moratoria in various countries – now finds approval with 58 per cent of respondents. And the solid support for medical applications of biotechnology spreads over to non-therapeutic applications. Moving from repair to improvement, we find that 56 per cent of the European public approves of research that aims to enhance human performance. However, support for regenerative medicine is not unconditional. Approval is contingent upon perceptions of adequate oversight and control.

Governance of science

Europeans' views on the governance of science were sought in the context of two examples of biotechnology: synthetic biology and animal cloning for food products. Respondents were asked to choose between, firstly, decision-making based on scientific evidence or on moral and ethical criteria, and secondly, decisions made on expert evidence or reflecting the views of the public. 52 per cent of European citizens believe that synthetic biology should be governed on the basis of scientific delegation where experts, not the public decide, and where evidence relating to risks and benefits, not moral concerns, are the key considerations. However, nearly a quarter of Europeans take the opposite view: it is the public, not experts, and moral concerns, not risks and benefits, that should dictate the principles of governance for such technologies (the principle of 'moral deliberation'). For animal cloning (compared to synthetic biology) some ten per cent fewer opt for scientific deliberation and nine per cent more opt for moral deliberation. It seems that moral and ethical issues are more salient for animal cloning for food products than for synthetic biology: altogether, 38 per cent of respondents choose a position prioritising moral and ethical issues for synthetic biology, with 49 per cent doing the same for animal cloning for food. To put this another way, the European public is evenly split between those viewing animal cloning for food as a moral issue and those viewing it as a scientific issue.

Public ethics, technological optimism and support for biotechnologies

Analysing the range of questions in the survey that address issues of public ethics – the moral and ethical issues raised by biotechnology and the life sciences – we find five clusters of countries. Key contrasts emerge between clusters of countries. First, those that prioritise science over ethics and those that prioritise ethics over science, and second those countries that are concerned about distributional fairness and those who are not. In combination these contrasts are related to people's optimism about the contribution of technologies to improving our way of life and support for regenerative medicines and other applications of biotechnology and the life sciences. Where ethics takes priority over science, concerns about distributional fairness lead to a profile of lower support; but in the absence of sensitivities about distributional fairness, the profile of support is relatively higher. When science taking priority over ethics is combined with concerns about distributional fairness, then we find only moderate support; but here again the absence of sensitivities about distributional fairness reveals a profile of high support.

⇨ *Europeans and Biotechnology in 2010: Winds of change?* http://ec.europa.eu/public_opinion/archives/ebs/ebs_341_winds_en.pdf

© *European Union, 1995-2011*

Chapter 2: ANIMAL CLONING

Animal cloning

Information from Understanding Animal Research.

Dolly the sheep may have been the world's most famous clone, but she was not the first. Cloning creates a genetically identical copy of an animal or plant. Many animals – including frogs, mice, sheep and cows – had been cloned before Dolly. Plants are often cloned – when you take a cutting, you are producing a clone. Human identical twins are also clones.

So Dolly was not the first clone, and she looked like any other sheep, so why did she cause so much excitement and concern? Because she was the first mammal to be cloned from an adult cell, rather than an embryo. This was a major scientific achievement, but also raised ethical concerns.

Since 1996, when Dolly was born, other sheep have been cloned from adult cells, as have mice, rabbits, horses and donkeys, pigs, goats and cattle. In 2004, a mouse was cloned using a nucleus from an olfactory neuron, showing that the donor nucleus can come from a tissue of the body that does not normally divide.

How was Dolly produced?

Producing an animal clone from an adult cell is obviously much more complex and difficult than growing a plant from a cutting. So when scientists working at the Roslin Institute in Scotland produced Dolly, the only lamb born from 277 attempts, it was a major news story around the world.

To produce Dolly, the scientists used the nucleus of an udder cell from a six-year-old Finn Dorset white sheep. The nucleus contains nearly all the cell's genes. They had to find a way to 'reprogramme' the udder cells – to keep them alive but stop them growing – which they achieved by altering the growth medium (the 'soup' in which the cells were kept alive). Then they injected the cell into an unfertilised egg cell which had had its nucleus removed, and made the cells fuse by using electrical pulses. The unfertilised egg cell came from a Scottish Blackface ewe.

When the scientists had managed to fuse the nucleus from the adult white sheep cell with the egg cell from the black-faced sheep, they needed to make sure that the resulting cell would develop into an embryo. They cultured it for six or seven days to see if it divided and developed normally, before implanting it into a surrogate mother, another Scottish Blackface ewe. Dolly had a white face.

To produce Dolly, the scientists used the nucleus of an udder cell from a six-year-old Finn Dorset white sheep

From 277 cell fusions, 29 early embryos developed and were implanted into 13 surrogate mothers. But only one pregnancy went to full term, and the 6.6kg Finn Dorset lamb 6LLS (alias Dolly) was born after 148 days.

Why are scientists interested in cloning?

The main reason that the scientists at Roslin wanted to be able to clone sheep and other large animals was connected with their research aimed at producing medicines in the milk of such animals. Researchers have managed to transfer human genes that produce useful proteins into sheep and cows, so that they can produce, for instance, the blood clotting agent factor IX to treat haemophilia or alpha-1-antitrypsin to treat cystic fibrosis and other lung conditions.

Cloned animals could also be developed that would produce human antibodies against infectious diseases and even cancers. 'Foreign' genes have been transplanted into zebra fish, which are widely used in laboratories, and embryos cloned from these fish express the foreign protein. If this technique can be applied to mammalian cells and the cells cultured to produce cloned animals, these could then breed conventionally to form flocks of genetically engineered animals all producing medicines in their milk.

There are other medical and scientific reasons for the interest in cloning. It is already being used alongside genetic techniques in the development of animal organs for transplant into humans (xenotransplantation). Combining such genetic techniques with cloning of pigs (achieved for the first time in March 2000) would lead

to a reliable supply of suitable donor organs. The use of pig organs has been hampered by the presence of a sugar, alpha gal, on pig cells, but in 2002, scientists succeeded in knocking out the gene that makes it, and these 'knockout' pigs could be bred naturally. However, there are still worries about virus transmission.

The study of animal clones and cloned cells could lead to greater understanding of the development of the embryo and of ageing and age-related diseases. Cloned mice become obese, with related symptoms such as raised plasma insulin and leptin levels, though their offspring do not and are normal. Cloning could be used to create better animal models of diseases, which could in turn lead to further progress in understanding and treating those diseases. It could even enhance biodiversity by ensuring the continuation of rare breeds and endangered species.

What happened to Dolly?

Dolly, probably the most famous sheep in the world, lived a pampered existence at the Roslin Institute. She mated and produced normal offspring in the normal way, showing that such cloned animals can reproduce. Born on 5 July 1996, she was euthanased on 14 February 2003, aged six and a half. Sheep can live to age 11 or 12, but Dolly suffered from arthritis in a hind leg joint and from sheep pulmonary adenomatosis, a virus-induced lung tumour to which sheep raised indoors are prone. On 2 February 2003, Australia's first cloned sheep died unexpectedly at the age of two years and ten months. The cause of death was unknown and the carcass was quickly cremated as it was decomposing.

Dolly's chromosomes were a little shorter than those of other sheep, but in most other ways she was the same as any other sheep of her chronological age. However, her early ageing may reflect that she was raised from the nucleus of a six-year-old sheep. Study of her cells also revealed that the very small amount of DNA outside the nucleus, in the mitochondria of the cells, is all inherited from the donor egg cell, not from the donor nucleus like the rest of her DNA. So she is not a completely identical copy. This finding could be important for sex-linked diseases such as haemophilia, and certain neuromuscular, brain and kidney conditions that are passed on through the mother's side of the family only.

Improving the technology

Scientists are working on ways to improve the technology. For example, when two genetically identical cloned mice embryos are combined, the aggregate embryo is more likely to survive to birth. Improvements in the culture medium may also help.

Ethical concerns and regulation

Most of the ethical concerns about cloning relate to the possibility that it might be used to clone humans. There would be enormous technical difficulties. As the technology stands at present, it would have to involve women willing to donate perhaps hundreds of eggs, surrogate pregnancies with high rates of miscarriage and stillbirth, and the possibility of premature ageing and high cancer rates for any children so produced. However, in 2004, South Korean scientists announced that they had cloned 30 human embryos, grown them in the laboratory until they were a hollow ball of cells, and produced a line of stem cells from them. Further ethical discussion was raised in 2008 when scientists succeeded in cloning mice from tissue that had been frozen for 16 years.

In the US, President Clinton asked the National Bioethics Commission and Congress to examine the issues, and in the UK the House of Commons Science and Technology Committee, the Human Embryology and Fertilisation Authority and the Human Genetics Advisory Commission all consulted widely and advised that human

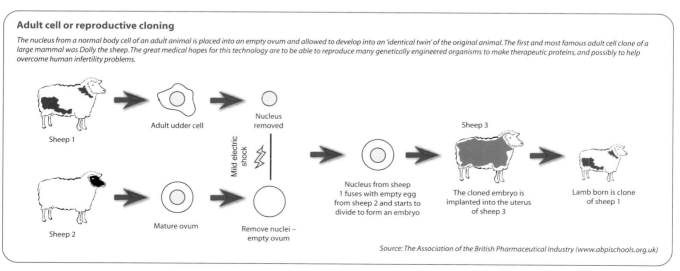

Adult cell or reproductive cloning

The nucleus from a normal body cell of an adult animal is placed into an empty ovum and allowed to develop into an 'identical twin' of the original animal. The first and most famous adult cell clone of a large mammal was Dolly the sheep. The great medical hopes for this technology are to be able to reproduce many genetically engineered organisms to make therapeutic proteins, and possibly to help overcome human infertility problems.

Source: The Association of the British Pharmaceutical Industry (www.abpischools.org.uk)

cloning should be banned. The Council of Europe has banned human cloning: in fact most countries have banned the use of cloning to produce human babies (human reproductive cloning). However, there is one important medical aspect of cloning technology that could be applied to humans, which people may find less objectionable. This is therapeutic cloning (or cell nucleus replacement) for tissue engineering, in which tissues, rather than a baby, are created.

All cloning for research or medical purposes in the UK must be approved by the Home Office under the strict controls of the Animals (Scientific Procedures) Act 1986

In therapeutic cloning, single cells would be taken from a person and 'reprogrammed' to create stem cells, which have the potential to develop into any type of cell in the body. When needed, the stem cells could be thawed and then induced to grow into particular types of cell such as heart, liver or brain cells that could be used in medical treatment. Reprogramming cells is likely to prove technically difficult.

Therapeutic cloning research is already being conducted in animals, and stem cells have been grown by this method and transplanted back into the original donor animal. In humans, this technique would revolutionise cell and tissue transplantation as a method of treating diseases. However, it is a very new science and has raised ethical concerns. In the UK, a group headed by the Chief Medical Officer, Professor Liam Donaldson, has recommended that research on early human embryos should be allowed. The Human Fertilisation and Embryology Act was amended in 2001 to allow the use of embryos for stem cell research and consequently the HFEA has the responsibility for regulating all embryonic stem cell research in the UK. There is a potential supply of early embryos as patients undergoing in-vitro fertilisation usually produce a surplus of fertilised eggs.

As far as animal cloning is concerned, all cloning for research or medical purposes in the UK must be approved by the Home Office under the strict controls of the Animals (Scientific Procedures) Act 1986. This safeguards animal welfare while allowing important scientific and medical research to go ahead.

⇨ The above information is reprinted with kind permission from Understanding Animal Research. Visit www.understandinganimalresearch.org.uk for more information.

© Understanding Animal Research

Timeline of domestic species cloned

⇨ **1996 – Dolly the sheep** first mammal cloned from an adult cell (Wilmut et al 1997)
⇨ **1998 – Cow** (Kato et al 1998)
⇨ **1998 – Mouse** (Wakayama et al 1998)
⇨ **1999 – Goat** (Baguisi et al 1999)
⇨ **2000 – Pig** (Polejaeva et al 2000)
⇨ **2002 – Rabbit** (Chesne et al 2002)
⇨ **2002 – Cat** (Shin et al 2002)
⇨ **2002 – Rat** (Zhou et al 2003)
⇨ **2003 – Horse** (Galli et al 2003)
⇨ **2003 – Mule** (Woods et al 2003)
⇨ **2003 – Deer** (Anon. 2003)
⇨ **2004 – Buffalo** (Shi et al 2007)
⇨ **2005 – Dog** (Lee et al 2005)
⇨ **2006 – Ferret** (Li et al 2006)
⇨ **2009 – Camel** (Wani et al 2010)

⇨ The above information is an extract from the Compassion in World Farming report *Farm Animal Cloning*, and is reprinted with permission. Visit www.ciwf.org.uk for more information and www.ciwf.org/education for educational materials.

© Compassion in World Farming

Q&A on cloning of animals for food

Information from Compassion in World Farming.

What is cloning?

The aim of cloning is to produce genetically identical copies of an animal.

Cloning involves collecting a cell from the animal that is to be cloned (called the 'donor cell') and transferring it into an egg cell that has been removed from another animal. The donor cell and the egg cell are fused by an electrical pulse and from this a cloned embryo is developed.

Once a cloned embryo has been produced, it is implanted into a surrogate (substitute) mother who carries out the pregnancy. This is an invasive process. In pigs the transfer of the embryo into the surrogate mother is performed by a surgical procedure. In cattle, embryo transfer is sufficiently stressful for UK law to require a general or epidural anaesthetic.

Why is cloning an animal welfare issue?

Scientific research shows cloning often involves severe suffering both for the surrogate mothers and for the clones themselves.

Painful births

Cloned calves tend to be heavier than normal which leads to painful births for the surrogate mothers and as a result Caesarean sections are often needed.

High mortality rates for clones

Most clones die during pregnancy. Of those that survive, many (up to 35%) die during or shortly after birth or in the early weeks of life from a range of problems including heart failure, respiratory difficulties, muscle and joint problems and defective immune systems.

The European Food Safety Authority (EFSA) has said that: 'The health and welfare of a significant proportion of clones have been found to be adversely affected, often severely and with a fatal outcome.'

Is cloning animals for food ethically justifiable?

The Opinion of the European Group on Ethics (EGE) in Science and New Technologies concluded that 'considering the current level of suffering and health problems of surrogate dams and animal clones, the EGE has doubts as to whether cloning animals for food supply is ethically justified'. The EGE added that it 'does not see convincing arguments to justify the production of food from clones and their offspring'.

In what ways is cloning likely to be used within the livestock sector?

Cloning aims to produce multiple copies of the highest yielding cows and fastest growing pigs. Traditional selective breeding has already led to major health problems for such animals. EFSA has concluded that 'genetic selection for high milk yield is the major factor causing poor welfare in dairy cows' and that genetic selection of pigs for rapid growth has led to leg and heart disorders. These animals are 'bred to suffer' – they are being pushed to their physical limits and often break down as a result. Cloning will exacerbate these problems. Cloning of the most fast-growing and high-yielding animals will lead to an even higher proportion of animals suffering from serious health and welfare problems.

How many clones are alive worldwide?

EFSA estimates that in 2007 there were about 100 cattle clones and fewer pig clones in the EU. The estimated number in the US is about 570 cattle and ten pig clones. There are also clones produced elsewhere, e.g. Argentina, Australia, China, Japan and New Zealand.

What is the situation in the UK?

Eight calves with a cloned mother have been born in the UK. The cloned cow was produced in the US and frozen embryos from her were imported by the UK. The Food Standards Agency (FSA) is investigating whether milk from one of the calves has entered the food chain. The FSA has reported that two of the male offspring of the cloned cow have been slaughtered. The first was slaughtered in 2009. Meat from this animal entered the food chain and will have been eaten. The second was slaughtered on 27 July 2010. Meat from this animal has been stopped from entering the food chain.

What is the position in the US?

Initially there was a voluntary moratorium in the US that prevented the sale of food from clones and their offspring. However, the moratorium has been lifted for meat and milk from the offspring of clones and we suspect that before long it will be lifted for food from the clones themselves.

What is the legal position?

Cloning is arguably unlawful under paragraph 20 of the Annex to EU Directive 98/58 which provides that: 'Natural or artificial breeding or breeding procedures which cause or are likely to cause suffering or injury to any of the animals concerned must not be practised.' However, the position should be clarified by passing a law that clearly bans the cloning of animals for food production. The Lisbon Treaty requires the EU and the Member States, when formulating and implementing their policies on agriculture, to 'pay full regard to the welfare requirements of animals'. If the EU does not ban cloning it will be in breach of its Treaty obligation to pay full regard to animals' welfare needs. The sale of meat and milk from clones is governed by the EU Novel Foods Regulation. We agree with the FSA that under the Regulation an authorisation must be granted before food from clones or their offspring can be put on sale. No such authorisation has been applied for in the UK.

A herd of cloned animals or their offspring will have less genetic diversity than conventional animals

Do we need cloning to feed the world?

No, we do not. Cloning is an inefficient way of feeding the growing world population. Clones and other high-yielding animals are fed on cereals. These crops could feed more people if they were used for direct human consumption rather than being fed to animals. The most efficient way of rearing cattle is to let them graze at pasture eating grass (with perhaps just a little supplementary feed). This way they are converting something we cannot consume, grass, into meat and milk that we can eat.

How will consumers react to milk and meat from cloned animals and their offspring?

UK and other EU consumers and supermarkets have rejected GM crops. They are likely to be even more uncomfortable with cloning as this interferes with the genetic makeup of animals. Indeed, opposition to cloning may be even greater as consumers may well feel that such a high-tech approach to sentient beings is even more disturbing than the genetic manipulation of crops. It would be unwise for farmers to use cloned animals and their offspring as this could tarnish the image of agriculture.

A recent Eurobarometer study found that:

⇨ 69% of interviewees agreed that animal cloning would risk treating animals as commodities rather than creatures with feelings.

⇨ 61% of EU citizens think animal cloning is morally wrong.

⇨ A majority (58%) of EU citizens are not willing to accept animal cloning for food production.

⇨ A majority said they were unlikely to buy meat or milk from cloned animals even if a trusted source stated that such products were safe to eat.

Is there a link between cloning and genetic engineering?

Animals are being genetically engineered in research facilities to increase the animals' rate of growth. Pigs, sheep and salmon engineered with growth hormone genes have been born with gross skeletal deformities and enlarged organs. Many die shortly before or after birth. Cloning is not genetic engineering. However, cloning is attractive to the biotechnology industry as, once a genetically engineered animal has been produced, the

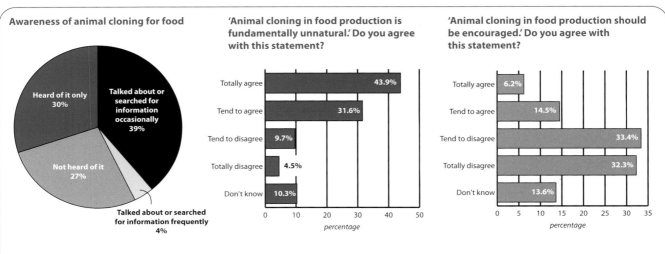

Source: Europeans and Biotechnology in 2010 Winds of change? http://ec.europa.eu, © European Union, 1995-2011

simplest way of making multiple copies of that animal is by cloning. Cloning will facilitate the genetic engineering of animals, a process that leads to great suffering.

What will be the implications of having herds of genetically very similar animals?

A herd of cloned animals or their offspring will have less genetic diversity than conventional animals and so will be much more vulnerable to any disease that enters the herd.

What are the wider implications of cloning?

An increasing number of people want society to move away from the cruelties of factory farming towards a more sustainable, humane agriculture. Cloning is taking us in the wrong direction – towards perpetuating industrial farming when all other societal trends point towards sustainable farming and respect for animals as sentient beings.

What is Compassion in World Farming's objective?

We want the EU to ban the cloning of animals for food and the sale of meat and dairy products from clones and their offspring.

What is the political position?

The EU is in the process of discussing a proposed new Novel Foods Regulation. In July 2010 the European Parliament voted for a ban on the sale of meat or dairy products derived from cloned animals or their offspring.

We agree with the EU Agriculture Council (the Agriculture Ministers from the 27 EU countries) that the proposed Novel Foods Regulation cannot address all the issues of cloning and that therefore the European Commission should produce a report on all aspects of food from cloned animals and their offspring. However, we disagree with the Council that in the meantime the Novel Foods Regulation should allow such food to be placed on the market.

Autumn 2010 is going to be a very active time politically in Europe. In September the Parliament and the Council will be holding talks to try and resolve their differences. The Commission is due to publish their report before the end of the year and this will generate a huge debate on whether the EU should permit the cloning of animals for food.

August 2010

⇨ The above information is reprinted with kind permission from Compassion in World Farming. Visit www.ciwf.org.uk for more information and www.ciwf.org/education for educational materials.

© *Compassion in World Farming*

Cloned cows

Nearly half (45%) of the British public say that they would not personally be happy to eat or drink food products from a cloned animal, compared to only 34% who would, a poll has discovered.

The results come after claims emerged this week that milk is on sale in Britain from cloned cows. Using cloned cows for meat or milk in the UK would be illegal without permission from the European Food Safety Authority, and currently there have not been any applications. However, in the US it has been cleared as safe to eat.

Interestingly, women seem more opposed to eating cloned meat, with 52% saying that they would not be happy to eat food from cloned animals, compared to 39% of men. In fact, turning the results on their head, almost double the amount of men (45%) said that they would be happy to eat such meat compared to women (23%).

Younger adults seem to be less worried about the origins of their meat, as the over-60s are more against the suggestion of using cloned animals, with 45% actively objecting compared to 39% of 18- to 24-year-olds. In fact, 18- to 24-year-olds are completely split, with 39% also feeling happy to personally condone the practice.

Cloning animals has been a much-debated topic since the birth of Dolly the sheep in 1996. The subject will surely continue to elicit strong emotions, as it remains to be seen whether animals like Dolly will one day become acceptable sources of food for the British public.

4 August 2010

⇨ The above information is reprinted with kind permission from YouGov. Visit http://today.yougov.co.uk for more information.

© *YouGov*

'If anything, this milk will be better quality'

In the wake of the NYT revelations, a genetics expert tells spiked that it's foolish to cry over cloned cows' milk.

By Tim Black

Following an article published in the *New York Times* on Friday, in which a UK-based dairy farmer admitted that his farm has been 'using milk from a cow bred from a clone as part of his daily production', the British media gradually picked up the bovine scent. Come Monday morning, the *Daily Mail* clearly felt it had a man-bites-dog sensation on its hands. 'Clone farm's milk is on sale', screamed its front page. While other media outlets have not been quite as titillated by this story of cows-make-milk as the *Mail* has been, it has still been causing a bit of a stir.

> **While the quality of the milk might be a question of taste, one thing that does seem certain is that there is no reason for thinking that this milk poses any health threat**

All of which is a little puzzling given that it's difficult to work out what exactly the problem with this cloned cow's offspring's milk actually is. As Dr Robin Lovell-Badge, head of developmental genetics at the UK National Institute for Medical Research, told spiked, 'One could argue that the milk from these cows is, if anything, actually going to be of a higher quality [than other cow breed's milk].'

While the quality of the milk might be a question of taste, one thing that does seem certain is that there is no reason for thinking that this milk poses any health threat. In the words of the US Food and Drug Administration in 2008, '[there is no known] science-based reason to distinguish between products from clones and products from conventionally produced animals'. In fact, insofar as these cows are merely the offspring of cloned cows – and not the cloned cow itself – they are essentially just like any other cow. 'There's no genetic modification at all,' explained Lovell-Badge. 'It's just like making an identical twin. You're not introducing any genetic traits.' In an attempt to put any health threat into further perspective, one geneticist told a *Telegraph* journalist that 'the chances would roughly be the same as that of your mother developing toxic milk while breastfeeding you as an infant'. That is, slim to non-existent.

The benefits of using cloned cows – whether the clones themselves or their offspring – are far from non-existent, however. And, in terms of rationale, neither is the procedure very new. Rather, cloning simply allows farmers to improve what they have been doing for thousands of years: breed their livestock selectively. The cows we see munching their way around the farm fields of the UK are not the amorous result of spontaneous bovine self-herding, but are the products of millennial-long human intervention. There has never been anything natural about it. As Lovell-Badge points out, 'Humans have been selectively breeding cattle for thousands of years. The various well-established breeds in the UK, for example, have been bred because they like the particular climate and the particular grass that we have here and they vary in terms of whether they're good for milk or beef. That is a product of selective breeding. The use of the cloning procedure is just a way of speeding up the selection of traits which are going to be beneficial.'

So, for example, having identified a cow with a high milk-yield, or one particularly resilient in a harsh, wet climate, the farmer can now call on cloning procedures to make an identical twin which will have the same properties. Whereas good old-fashioned selective breeding could take ages to fix a trait in a herd of cattle, cloning in this way is far more rapid and reliable.

> **Cloning simply allows farmers to improve what they have been doing for thousands of years: breed their livestock selectively**

With the benefits eclipsing any mythical health threat, it does raise the question: what exactly is everyone's problem? Ostensibly, it's a legal thing. Under existing European Union legislation, one is not allowed to market a 'novel food' – 'a food or food ingredient that does not have a significant history of consumption within the EU before 15 May 1997' – until it has been officially approved. The UK Food Standards Agency says that no such approval has been given, hence the *New York Times*' dairy farmer is in trouble.

But there is more to this than a regulation breach. This crying over cloned milk testifies to a state-endorsed suspicion of genetic modification and cloning *per se*. Hence, while the existing EU legislation permits 'novel food', official approval permitting, a bill recently passed by the European Parliament will, when implemented, lead to a blanket ban on meat and milk from clones and their offspring. A French MEP, Corinne Lepage, explained the decision: 'Although no safety concerns have been identified so far with meat produced from cloned animals, this technique raises serious issues about animal welfare, reduction of biodiversity, as well as ethical concerns.'

What is clear from Lepage's statement is that the actual foodstuffs produced, no matter how indirectly, from cloned animals are not the problem. Rather, the issues concerning Lepage lie outside the scientific endeavour, whether in vague 'ethical concerns' or in so-called animal welfare. Moreover, these concerns themselves need neither be fully articulated nor valid. What counts is simply the allusion to them, the suggestion that these are concerns that exist among the general public. In the UK a co-author of an FSA report on clone farming and food used the same tactic to justify the UK's regulation-heavy approach: 'The majority of people [questioned] came to the conclusion that they would not want to eat such food. The overwhelming majority either did not want it or were unsure.' Likewise, Peter Stevenson of Compassion in World Farming says: 'I would be appalled if milk from a clone offspring cow is coming into the food chain in the UK… the public is deeply concerned about this and does not want it.'

The concerns are not groundless, of course. Take the issue of 'animal welfare': 'Clearly when doing the cloning procedure, you start off with quite a few attempts before you get an embryo that is likely to develop', says Lovell-Badge. 'And then some embryos will fail during pregnancy, and then some that will be born may carry some abnormalities. So there are animal welfare issues here.'

'But', he adds, 'cattle, as it turns out, are not as affected as other species are by the cloning procedure, so it's a lot more efficient and there's far less evidence of abnormalities. After mice, cows are probably the species that have been cloned the most, so there's a lot of experience in cloning cows.'

What soon becomes clear is that the Europe-wide clampdown on cloned animal food derives not from any tangible health risk, or from the sadistic maltreatment of animals, but from officialdom's perception of what the public fears and thinks, no matter how irrational. And no wonder the public might seem anxious about such new genetic procedures and technologies. In this area, the state has for far too long seemed intent, not on dispelling fears and concerns, but on simply managing them. Happy to invoke the science when it suits a particular policy objective, the UK Government and others have repeatedly proven feeble and cowardly when dealing with issues thrown up by the actual science – especially genetics. So, before the disproportionately vocal objections of animal welfare groups, the state has lacked the authority to stick up, not for animals, but for those for whom farmers actually raise livestock: namely, the human species. And faced by public fears of the possible effects of fiddling with nature, the state has lacked the confidence to defend scientific experimentation.

> **This crying over cloned milk testifies to a state-endorsed suspicion of genetic modification and cloning**

It is this weakness, this lack of authority, that seems to lie at the source of the farcical furore around cloned cows' milk. Too afraid of the presumed public reaction to cloning technologies, a weak state draws on it instead as a source of legitimacy.

3 August 2010

⇨ The above information is reprinted with kind permission from spiked. Visit www.spiked-online.com for more information.

© *spiked*

Food watchdog admits tracking cloned cows 'impossible'

Britain's food safety watchdog has admitted it doesn't know how many cloned embryos have entered Britain after meat produced from cloned cows ended up in food.

By Harry Wallop, Bruno Waterfield in Brussels and Alastair Jamieson

Tim Smith, chief executive of the Food Standards Agency, said it was unsure if any other cloned embryos had been imported and conceded there was disagreement with Brussels over interpretation of the law.

However, he insisted the FSA had 'no concerns' about the safety of meat or milk from animals derived from clones but said a 'live investigation' was underway into how a three-year-old bull derived from a cloned animal was slaughtered, sold and eaten last year.

Under European law, foodstuffs, including milk, produced from cloned animals must pass a safety evaluation and gain authorisation before they are marketed

It is thought the meat, likely to have ended up in a pie or burger, came from a farm in Scotland.

Its admission came despite two days of strong denials from both the FSA and the dairy industry that no milk or meat from cloned animals or their offspring had ever entered the food chain.

Under European law, foodstuffs – including milk – produced from cloned animals must pass a safety evaluation and gain authorisation before they are marketed. But the FSA said it had neither made any authorisations nor been asked to do so.

The RSPCA said EU regulations gave 'mixed messages' to farmers while European officials said it was 'probable' thousands of cheese and meat products on sale in British supermarkets had come from animals that were derived from clones.

Mr Smith said it would be 'the autumn' before 'clarification' emerged on the law and that it was not yet clear what action, if any, could be taken against farmers who have traded cloned cattle or their offspring.

He told the BBC Radio 4 Today programme: 'There's a live investigation going on at the moment and, whilst we have got a first-class cattle tracing scheme, what we don't know is precisely how many embryos have been imported into the country.'

He said such a situation was 'inevitable', adding: 'It's a bit like the police being there and us expecting no crime. However good the system is, it relies on the honesty of those people participating.

'It's impossible for us to stand by each animal and watch it through each phase of its life cycle.'

David Bowles, a spokesman for the RSPCA, which opposes cloning, said: 'The onus is on producers to declare whether any foodstuffs come from clones or their offspring. Unless things are tightened up with a total ban, there is a certain inevitability of these foodstuffs leaking into the UK and into the food chain.'

Food from a cloned animal, or the offspring of a cloned animal, is illegal, according to the FSA. European officials – in charge of setting all British food safety laws – have said it is 'probable' that thousands of cheese and meat products on sale in British supermarkets had come from animals that were derived from clones.

An EU official said that because there were no restrictions on importing semen which had come from a cloned animal, it was possible that thousands of pigs and cows in Europe were the offspring of cloned animals. Millions of shots of semen are imported into the UK every year, and the Department for Agriculture confirmed it did not monitor whether they were from cloned animals or not.

There is no evidence that eating milk or meat from an animal that has been cloned or its offspring is detrimental to human health, but animal welfare campaigners argue the cloning process is very harmful for the animals themselves and produces sheep and cows with a shortened lifespan.

The row over cloning erupted after an unnamed farmer claimed in an American newspaper that he had illegally sold milk from the offspring of a cloned cow.

This has yet to be proved, but Holstein UK, responsible for registering all pedigree cows and bulls on farms, has confirmed that there are 97 calves in a herd in Scotland, which are all derived from the offspring of a cloned cow.

All the animals ultimately come from Vandyk-K Integ Paradise 2, a cloned cow in America. She was cloned using cells from the ear of a Holstein, a milking cow, about five years ago.

Vandyk-K was artificially inseminated with semen from a prize-winning bull in America. Embryos were then removed from her, frozen and flown to the UK, where they were implanted in two surrogate mother cows.

Her offspring include Dundee Perfect and Dundee Paratrooper. The FSA said yesterday that Perfect was slaughtered only last week and the meat had been intercepted before entering the food chain.

Dundee Paratrooper was slaughtered a year ago. 'Meat from this animal entered the food chain and will have been eaten,' the FSA said.

Molly Conisbee, of the Soil Association, which campaigns for organic food, said: 'This is a huge matter of concern. We were assured by the regulators that no cloned food had entered the food chain, and now it appears it has. There has been a huge failure of the system, of regulation and of labelling.'

The two bulls, now slaughtered, sired 97 cows. They are currently too young to be producing milk, it is believed. But the FSA is checking whether this is the case and whether any other offspring of Vandyk-K have entered the food chain.

Officials in Brussels, however, suggested that it could be far more than just the one bull that had entered the food chain.

An EU official said: 'It is a little ridiculous to talk about these embryos because semen trading is so huge.

'We import millions of doses of semen from the US and Canada and it is more than probable among them are doses from cloned animals. If just one per cent or 0.1 per cent is from cloned animals then there are hundreds or thousands of first-generation offspring. Second generation bred from the first would be an even greater number. The issue is exactly the same for pigs.'

'If the FSA then reached the likely conclusion that hundreds or thousands of British cows are offspring of cloned animals what will they do? Destroy the milk or slaughter the animals?'

Holstein UK said that though Defra did not monitor whether the semen was from a cloned animal, it did keep a record, and to its knowledge no semen from cloned animals had been imported into Britain.

However, officials in Brussels suggested it was possible that European food imported to Britain, such as salami and French or Italian cheese, could well be ultimately derived from cloned animals, and no label or restrictions could stop that flow of trade.

Government figures indicated that Britain imported £900 million worth of cheese last year.

4 August 2010

© Telegraph Media Group Limited 2010

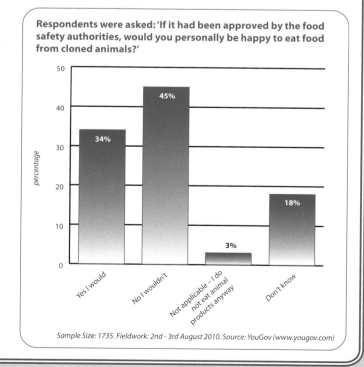

Respondents were asked: 'If it had been approved by the food safety authorities, would you personally be happy to eat food from cloned animals?'

- Yes I would: 34%
- No I wouldn't: 45%
- Not applicable – I do not eat animal products anyway: 3%
- Don't know: 18%

Sample Size: 1735. Fieldwork: 2nd - 3rd August 2010. Source: YouGov (www.yougov.com)

Why I'd happily eat lab-grown meat

Laboratory-grown meat is getting closer to reality – and we needn't be repulsed, says Leo Hickman.

What's the most disgusting thing you've ever put in your mouth? Never mind the kangaroo testicles and witchetty grubs pushed before jungle-strewn celebrities: earlier this year I met a man who has tasted a substance that seems to trigger the gag reflex in most people.

Willem van Eelen is an 86-year-old Dutchman who survived five years of near-starvation as a PoW in Japan to become the self-proclaimed 'Godfather of in vitro meat'. He now holds the patent to a food technology that repulses and excites people in equal measure. In his Amsterdam apartment he admitted to me, with a mischievous smile, that he once placed a small quantity of in vitro meat – muscle cells grown artificially, rather than in a living animal – in his mouth: 'I once put a few cells on the tip of my tongue. I couldn't resist it. It tasted a little like chicken.'

In vitro meat – also known as cultured or fake meat – is becoming a holy grail for anyone concerned about the environmental and ethical impacts of rearing millions of animals around the world each year for human consumption. Where today we use animals to turn grass into edible protein, in the future we might bypass this inefficient process and grow edible protein in an algae solution in factories instead. The animal rights group PETA has gone as far as offering a $1 million prize to the first scientists to bring the meat to market.

> **As a substitute for real meat – one that could boast environmental and animal-welfare positives – it seems too good to leave off the menu**

In what is an encouraging breakthrough, a team of Dutch scientists, with van Eelen as their figurehead, say they have now grown in vitro meat successfully for the first time in the laboratory.

When I visited Mark Post, one of the team's lead scientists, at his lab at the University of Technology in Eindhoven earlier this year, he showed me the plastic dishes in which the meat cells, originally taken from a pig foetus, were being grown. Inside was a pink liquid he described as having the texture of 'an undercooked egg'. It had yet to be 'exercised' via electrical stimulation into a muscle-type texture. Not exactly a pork chop with herb butter, but I said I was willing to taste the future.

Alas, he declined. Wait a few years, he said, and he should have got to the stage where long strips of it can be grown and rolled up into frankfurter-like sausages. It will probably be 30 years before we see in vitro pork chops, or anything more sophisticated than the processed meat found in cheap burgers and sausages.

On paper, the science sounds compelling, but it will take a superhuman effort on the part of the world's advertising agencies to convince us to swallow it. After living through various food scares – many involving meat – you can understand the hesitancy. The scientists admit all this, saying they've even thought of alternative names for it. ('Krea', the Greek word for meat, was one of their favourite suggestions.)

But I for one don't see a problem with placing a forkful of the stuff in my mouth. OK, it's never going to be a gourmet experience, but as a substitute for real meat – one that could boast environmental and animal-welfare positives – it seems too good to leave off the menu. Bon appetit.

1 December 2009

© Guardian News and Media Limited 2009

Chapter 3: HUMAN CLONING

Therapeutic cloning (somatic cell nuclear transfer)

Information from the Australian Stem Cell Centre.

What is therapeutic cloning?

Therapeutic cloning

Therapeutic cloning refers to the removal of a nucleus, which contains the genetic material, from virtually any cell of the body (a somatic cell) and its transfer by injection into an unfertilised egg from which the nucleus has also been removed. The newly reconstituted entity then starts dividing. After four to five days in culture, embryonic stem cells can then be removed and used to create many embryonic stem cells in culture. These embryonic stem cell 'lines' are genetically identical to the cell from which the DNA was originally removed.

Therapeutic cloning is also known as somatic cell nuclear transfer (SNCT) as the term cloning is frequently misunderstood by the general public. The word 'cloning' more often conjures up thoughts and beliefs about reproductive cloning.

> **Therapeutic cloning refers to the removal of a nucleus, which contains the genetic material, from virtually any cell of the body and its transfer by injection into an unfertilised egg from which the nucleus has been removed**

Reproductive cloning

Reproductive cloning is if a newly-formed embryo, resulting from a therapeutic cloning procedure, were transferred into the womb of a woman, it could theoretically, develop into a fetus. This technique has been used to clone agricultural animals, endangered species and recently domestic pets and primates, but has not been proven in humans. The scientific community overwhelmingly rejects the use of therapeutic cloning for the purposes of human reproductive cloning. Reproductive cloning is illegal in Australia and in many other countries.

While the procedure of therapeutic cloning employs aspects of cloning technology, researchers today are interested in therapeutic cloning as a means of deriving human embryonic stem cell lines for use in research and, ultimately, therapy. Therapeutic cloning has been legal in Australia since 2006 under the Prohibition of Human Cloning for Reproduction and the Regulation of Human Embryo Research Amendment Act 2006. Any Australian researchers wishing to use therapeutic cloning must apply to the National Health & Medical Research Council (NHMRC) for a licence.

> **Reproductive cloning is if a newly-formed embryo, resulting from a therapeutic cloning procedure, were transferred into the womb of a woman**

What is special about therapeutic cloning?

The capacity of therapeutic cloning to re-programme adult nuclei is extraordinary and unique. Cells of particular tissues generally express a characteristic set of genes, whether they are primitive stem cells, fully-differentiated (i.e. mature) cells, or something in between these extremes. Particularly for more mature cells, the tissue-specific patterns of gene expression are quite stable through many rounds of cell division.

Upon transfer to an enucleated egg, the adult nucleus becomes re-programmed in the environment of the egg. That is, genes that were not used before (switched off) become reactivated. A poorly-understood process, re-programming involves dramatic changes in the pattern of genes which are active in the nucleus. Instead of the adult nucleus causing the egg to behave like an adult cell, the egg causes the nucleus to go backwards along a

differentiation sequence, resulting in an embryonic-type cell. As a result of therapeutic cloning, the previously unfertilised egg takes on the properties of a fertilised egg and begins the first stages of development into an embryo.

What is known about the embryos resulting from therapeutic cloning?

In broad terms, embryos arising from therapeutic cloning are the same as embryos arising from fertilisation of an egg by a sperm. In agricultural research, therapeutic cloning has been used to create embryos in the laboratory which have been transferred into animals and given rise to offspring. 'Dolly' the sheep was born as a result of this procedure.

In broad terms, embryos arising from therapeutic cloning are the same as embryos arising from fertilisation of an egg by a sperm

However, there are also differences between a normally fertilised egg and one produced by therapeutic cloning, which are not generally understood. In animal studies clones appear to have increased abnormality and decreased pregnancy rates. For example, despite being relatively young for a sheep, which can live to 11 or 12 years of age, Dolly died prematurely at the age of six years, showing signs of arthritis and lung infection. Other than the major ethical concerns, this is a fundamental reason why reproductive cloning should not be performed on humans.

Research involving therapeutic cloning and its potential application

While the scientific community overwhelmingly rejects the use of therapeutic cloning for reproductive cloning, it would provide an invaluable tool for basic research.

As reported in *Nature* in November 2007, scientists successfully extracted stem cells from non-human cloned primate embryos. Although it is a highly significant achievement it must be considered that it took 304 eggs to produce two successful embryonic stem cell lines.

In January 2008 a Californian company, Stemagen, reported it had successfully cloned a human blastocyst, an early stage embryo. The Stemagen embryos were the first to be made with human adult skin cells through therapeutic cloning; however, they did not attempt to produce stem cell lines, focusing their attention on extensive genetic tests to prove the identity of the cloned embryos.

Researchers regard therapeutic cloning as an effective method for deriving human embryonic stem cells with specific characteristics, about which a great deal remains to be known and understood. The promise of therapeutic cloning is that it will be an effective way to derive embryonic stem cells which can then be used for the development of patient- and disease-specific cell-based therapies as well as the production of stem cells with specific disease characteristics for research purposes.

The use of a patient's own cells for tissue replacement through therapeutic cloning overcomes the problem of immune rejection that is a major complication of tissue transplantation today. Embryonic stem cells derived by nuclear transfer may, in the future, be used to treat diseases including diabetes, heart disease and Parkinson's disease.

July 2010

⇨ The above information is reprinted with kind permission from the Australian Stem Cell Centre. Visit www.stemcellcentre.edu.au for more information.

© *Australian Stem Cell Centre*

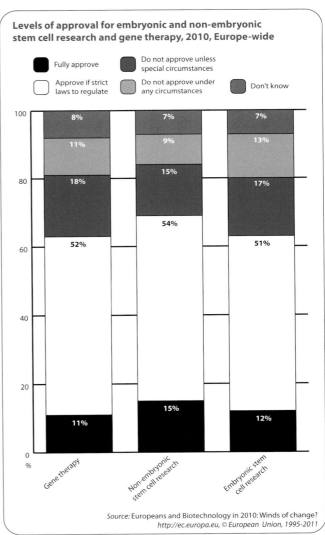

Levels of approval for embryonic and non-embryonic stem cell research and gene therapy, 2010, Europe-wide

Source: Europeans and Biotechnology in 2010: Winds of change? http://ec.europa.eu, © European Union, 1995-2011

Reproductive cloning

Science and policy history.

Human reproductive cloning is the creation of an individual who has identical nuclear genetic material (DNA) to an existing human being, and who is allowed to develop to term and beyond. Human reproductive cloning is widely regarded as unethical and inappropriate and is specifically prohibited in many jurisdictions.

An initiative at the United Nations General Assembly led to the United Nations Declaration on Human Cloning, which calls on all states to prohibit all forms of human cloning.

Basic scientific concepts

A gene is a hereditary unit consisting of a sequence of DNA that occupies a specific location on a chromosome. Chromosomes consist of long coiled chains of genes and are found within all nucleated cells in the human body. Human beings normally have 23 pairs of chromosomes; one of each pair is inherited from the genetic mother and one from the genetic father.

In sexual reproduction, a child receives half of their genes from the mother (contained in the egg) and half from the father (contained in the sperm). The combination of maternal and paternal genes which occurs at fertilisation forms the basis of human genetic variety and diversity. A small amount of genetic material is contained within mitochondria within the egg and this mitochondrial DNA is passed on to the child entirely from the mother.

In embryo splitting, an early human embryo divides into two genetically identical embryos which are then capable of developing independently. This process may happen spontaneously and is the mechanism whereby genetically identical twins (technically described as monozygotic twins) are formed. Embryo splitting can also be induced artificially.

In reproductive cloning the entire genetic code (except for the mitochondrial DNA) is reproduced from a single body cell of an adult individual. The most common cloning technique is somatic cell nuclear transfer (SCNT). The procedure is as follows.

1. The nucleus is removed from an egg, leaving the cytoplasm and mitochondria (cellular components derived from the mother).

2. A body (or somatic) cell is taken from an adult individual who is to be cloned. The DNA is extracted from the nucleus and inserted into the prepared egg.

3. The new cell is then induced to divide using either chemical or electrical stimulation, thereby commencing the development of an embryo.

4. After several days the dividing embryo is then placed into the womb of the recipient and allowed to develop to term. The result is a clone – an individual that is the genetic duplicate of the individual from whom the original body cell was taken. To date this process has not been proven to occur in a human being. If it did so, it is important to note that the resulting child would neither be the individual's son or daughter, nor their twin brother or sister. The child would truly be a new category of human being – a clone.

Hans Spemann, an eminent German scientist, conducted the first nuclear transfer experiment in 1928, in which he transferred the nucleus of a salamander embryo cell to a cell without a nucleus. This was followed up in 1952 with experiments carried out by Robert Briggs and Thomas J. King, using embryonic donor cell nuclei from amphibia. Earlier attempts by the two had failed, but by the completion of the project they had been successful in cloning 35 complete embryos and 27 tadpoles from 104 successful nuclear transfers. At the time they were unaware of the work of Hans Spemann. The work of Briggs and King helped initiate decades of substantial research into cloning.

John Gurdon of Oxford University discovered in 1962 that differentiated cell nuclei could also result in cloned offspring, the result of which proved that as differentiation occurred, there was no loss of genetic material. Whilst this attracted the attention of many, some scientists doubted the validity of Gurdon's work, uncovering flaws.

Research and development continued into nuclear transfer work in the use of mammals in the 1970s and all the way through to the 1990s, resulting in the conception of the first mammal cloned from an adult nucleus in 1996. The scientists behind this were by Ian Wilmut and Keith Campbell of the Roslin Institute, Scotland. The birth of the clone, Dolly the sheep, was first announced in the journal *Nature* in 1997 and initiated worldwide discussion about the possibility of cloning humans.

Since the birth of Dolly the sheep, there have been other species of animals cloned from differentiated cells, including horses, mice and cats.

These results all point to the possibility of somatic nuclear transfer eventually being successful in all mammals, including humans. Yet it is still the most commonly-held opinion that reproductive cloning of humans should be opposed.

Various claims have been made with regard to reproductive cloning but none of them have yet been able to offer documented evidence of such breakthroughs.

Consequently, observers treat such claims with great caution.

Policy

In 1984, the report of the Warnock Committee recommended that human cloning should not be permitted. However, current legislation in the UK concerning human cloning is somewhat ambiguous.

In 1990, the Human Fertilisation and Embryology (HFE) Act became law. This Act gave statutory force to the recommendations of the Warnock Report and set up the Human Fertilisation and Embryology Authority (HFEA) to oversee the Act's legislation and operation. The HFE Act allowed in vitro fertilisation along with embryo freezing, research and destruction. For the first time, the Act permitted licensed individuals to experiment with embryos during their first 14 days as well allowing consent to be obtained for embryos to be used in IVF treatment for infertile couples and to be donated for the infertility treatment of others.

In order to consider and explore the ethical issues and possible applications of human cloning, in 1998 the UK Government asked the HFEA and the Human Genetics Advisory Commission (now called the Human Genetics Commission) to undertake research into these issues and present a report on their findings. The conclusions of this project were published in the report *Cloning Issues in Reproduction, Science and Medicine*. The report included a recommendation stating that no human reproductive cloning should be allowed but that therapeutic cloning should be. In 1999, in answer to a question in the House of Lords, the official UK Government position was that 'human reproductive cloning is ethically unacceptable and cannot take place in this country'.

Whilst the law did not allow any form of reproductive cloning to be licensed, there was no explicit ban on cloning. The arrival of Dolly the sheep in 1996 caused worldwide fears to arise that the new technology could be applied to humans. In the UK, the HFEA responded by clearly stating that depending on the method used, cloning was either prohibited or subject to a licensing requirement.

Yet further analysis of the 1990 Act led many, including law and medical ethics academics Mason and McCall Smith, to be openly uncertain as to whether or not reproductive cloning was prohibited by law, due to the rather vague and ambiguous terminology used in the 1990 Act. The definition of the embryo used by the 1990 Act in s.1(1) is 'a live human embryo where fertilisation is complete', including 'an egg in the process of fertilisation'. Yet the view of certain scientists was that the Dolly technique did not involve an act of fertilisation.

Other academics also highlighted this potential regulatory gap, which was brought into the full media spotlight when Bruno Quintavalle brought a judicial review action concerning this question on behalf of the Pro-Life Alliance. As the review took place, in August 2000 an independent advisory group commissioned by the Government and chaired by the Chief Medical Officer Dr Liam Donaldson presented its recommendations on the future development and benefits of research or therapeutic cloning and reproductive cloning. Aside from its recommendations concerning research cloning, it recommended that reproductive cloning remained illegal in the UK.

The Government fully accepted the recommendations of the report, which led to the introduction of the Human Reproductive Cloning Act (passed on 4 December 2001). This outlaws any attempts at reproductive cloning. Section 1(1) declares that it is a criminal offence to place 'in a woman a human embryo which has been created otherwise than by fertilisation'. Furthermore, the HFEA has made it clear that, at least for the foreseeable future, a licence will not be issued for embryo splitting or any other form of reproductive cloning that falls within the remit of its activity.

In 2001, the Act of 1990 was subsequently amended by way of the Human Fertilisation and Embryology (Research Purposes) Regulations 2001. This allowed for the cloning of human embryos for research into diseases up until their 14th day.

Today, reproductive cloning is illegal in most countries. However, this does not deter some scientists from attempting to clone a human being. Such attempts are

not hampered by lack of resources or knowledge. Most assisted conceptions centres around the world possess most of the necessary equipment and facilities, and many professionals possess the knowledge and skills to carry out the procedures.

In April 2009, American fertility expert Dr. Panayiotis Zavos hit the headlines when he claimed that he had attempted to clone four individuals at a secret laboratory, as well as producing a cloned embryo of three people who had died. Zavos commented: 'There is absolutely no doubt about it, and I may not be the one that does it, but the cloned child is coming. There is absolutely no way that it will not happen…to date we have had over 100 enquiries and every enquiry is serious.'

Ethics

The ethical and social implications presented by reproductive cloning are profound, requiring careful reflection and consideration. To reduce these issues down to a series of arguments for and against can be a little too simplistic. Nevertheless, such an approach at least highlights some of the key questions which require our reflection at a deeper level.

Arguments against the creation of human clones

The majority of arguments against reproductive cloning have highlighted the possible adverse consequences on individuals, family relationships and society as a whole. However, principled objections to human cloning in itself have also been raised.

a) Principled objections to reproductive cloning

⇨ Instrumentalisation of human beings. Cloning represents the creation of a human individual as an instrument of another human's will and purposes. It reflects a view of humans as objects that can be tailor-designed and manufactured to meet certain characteristics and specifications.

⇨ The child as a reflection of the love between a man and a woman. In orthodox Christian thinking, human procreation is seen as indissolubly linked to the committed love of a man and a woman. In other words, 'making love' and 'making babies' belong together and every child should be a reflection of a love relationship. Reproductive cloning, as an asexual form of reproduction, destroys the link between a human relationship and the creation of a child.

⇨ Uniqueness diminished. Reproductive cloning threatens widely- and deeply-held convictions about the individuality of human beings, which are closely linked to notions of human freedom.

b) Adverse consequences of reproductive cloning

⇨ Safety fears concerning human cloning. Current scientific experience indicates that at least 95% of mammalian cloning experiments have resulted in miscarriages, still births and life threatening abnormalities. There is increasing evidence that mammalian clones are likely to have subtle abnormalities of complex and poorly-understood genetic control mechanisms. In 2001, at a National Academy of Sciences conference in Washington, DC, scientists who opposed reproductive human cloning presented the results of research, which indicated that approximately a third of cloned mammals have developed abnormalities. Such abnormalities include large offspring syndrome (LOS), where the offspring is born oversized with disproportionately large internal organs and suffers from respiratory, circulatory and other problems. Many argue that with current understanding of cloning techniques, similar abnormalities would probably occur in human somatic cell nuclear transfer.

- Social and psychological consequences of cloning. It has been argued that a child clone would inevitably suffer adversely from the existence of their nuclear donor and from the knowledge that they had been created for a specific purpose. Thus a 'replacement' child would suffer from continuous comparisons and memories of somebody else, and would not experience the normal celebrations of a new person, a new life: *their* life. Simply creating a clone to replace a beloved child tends to dehumanise both the child and its replacement clone. Similarly, a child that was created to be a genetically identical donor for an existing human may feel coerced or abused by the process. Reproductive clones may also encounter various forms of social discrimination and stigmatisation.

- The slippery slope argument. Once reproductive human cloning is permitted, it may become more difficult to prohibit and restrict other, more dangerous applications of genetic and reproductive technology. The technology can easily be utilised outside governmental scrutiny and is ultimately impossible to control.

Arguments in favour of creating human clones

- The existence of identical twins. It is argued that identical twins represent a natural form of cloning. However, whilst the identical twins share the same genetic make-up, they are not deliberately planned copies of their parents created by asexual reproduction. In addition, identical twins are the same age as one another and recognise each other as siblings, whereas a cloned child would be genetically the same as a parent or another human of a different age.

- A novel form of human reproduction. For those who are unable to have children through other fertility treatments (i.e. those who produce neither eggs nor sperm), human cloning may provide a viable alternative to having a genetically-related child. However, individuals who are in this position are relatively rare.

- Children for lesbian couples or single women. Human cloning would provide lesbian couples or single women with an opportunity to have a child without using donor sperm.

- My child: mark II. Protagonists of human cloning such as Dr. Richard Seed argue that parents of a child who died prematurely or even through a tragic accident would, through reproductive cloning, be able to compensate for their loss by having a 'second version' of their child. On the other hand, seeking to create a 'replacement child' may not bring all that the parents aspire to replace. Indeed, the clone will be exposed to different environmental influences and will develop in a different way from the original child.

- Creation of a genetically identical donor. Cloning will enable the creation of a genetically identical child who may provide desperately needed tissue or organs for the cloned adult which cannot be obtained in any other way. A clone could provide bone marrow, tissue (such as skin or muscle) or even solid organs such as a kidney.

- Safety concerns can be overcome. Proponents of cloning argue that most medical technology brings with it an element of risk and that normal human reproduction is not risk-free. There were initial safety concerns about in vitro fertilisation, but this is now widely accepted and practised. It is also argued that incidents of LOS are probably due to poor embryonic culture conditions. The techniques of embryo manipulation are rapidly improving and the problem may be eradicated. Once the process of reproductive cloning has been improved and perfected in mammals, it is believed that human trials can be commenced with a reasonable degree of safety.

- The slippery slope can be prevented. It is argued that robust legislation, democratic accountability and effective governance of professional practices will prevent the abuse of reproductive technology. Gradual extension of practices is not inevitable and as a society we have the mechanisms to prevent unacceptable developments from occurring.

- A reproductive right? Some believe that human cloning is a reproductive right: hence from a libertarian perspective it should be allowed without restriction, provided that the safety concerns have been overcome and cloning is demonstrated to be no less safe than natural reproduction. Yet rights are socially negotiated and confer duties and responsibilities on others.

Writing on his prophetic novel, Aldous Huxley said: 'The theme of *Brave New World* is not the advancement of science as such; it is the advancement of science as it affects human individuals.' The issues surrounding reproductive cloning are of crucial importance because they ask the essential questions: what does it mean to be human, and in light of this, what is the appropriate use of technology?

April 2011

⇨ The above information is reprinted with kind permission from BioCentre. Visit www.bioethics.ac.uk for more information.

© BioCentre

What are stem cells?

Stem cells are stirring up great excitement in medical research. What are they and why are scientists so intrigued by them?

What are stem cells?

The human body is made up of about 200 different kinds of specialised cells such as muscle cells, nerve cells, fat cells and skin cells. All specialised cells originate from stem cells. A stem cell is a cell that is not yet specialised. The process of specialisation is called differentiation and once the differentiation pathway of a stem cell has been decided, it can no longer become another type of cell.

Different types of stem cells have different levels of potential. A stem cell that can become every type of cell in the body is called pluripotent and a stem cell that can become only some types of cells is called multipotent.

Where are stem cells found?

Stem cells are found in the early embryo, the foetus, amniotic fluid, the placenta and umbilical cord blood. After birth and for the rest of life, stem cells continue to reside in many sites of the body, including skin, hair follicles, bone marrow and blood, brain and spinal cord, the lining of the nose, gut, lung, joint fluid, muscle, fat and menstrual blood, to name a few.

> *A stem cell that can become every type of cell in the body is called pluripotent and a stem cell that can become only some types of cell is called multipotent*

In the growing body, stem cells are responsible for generating new tissues, and once growth is complete, stem cells are responsible for repair and regeneration of damaged and ageing tissues.

Stem cells can be divided into two broad groups: tissue-specific stem cells (also known as adult stem cells) and pluripotent stem cells (including embryonic stem cells and iPS cells). Tissue-specific stem cells are derived from, or resident in, adult tissues, and can usually only give rise to the cells of that tissue, thus they are considered multipotent. Embryonic stem cells, derived from a small group of cells in the early embryo (five to seven days), are undifferentiated and are considered pluripotent as they can become every type of cell in the body. Recently, scientists discovered that a mature, fully-specialised cell, for example a human skin cell, in the right conditions could be induced to mimic the characteristics of an embryonic stem cell. These are known as induced pluripotent stem cells (iPS cells).

Why are stem cells so different?

Stem cells are different from other cells in the body in three main ways:

1. Stem cells are unspecialised. They have not developed into cells that perform a specific function.

> *In the growing body, stem cells are responsible for generating new tissues, and once growth is complete, stem cells are responsible for repair and regeneration of damaged and ageing tissues*

2. Stem cells can differentiate. This means they can divide and produce cells that have the potential to become other, more specific cell types, tissues or organs. These new cells and tissues are used to repair or replace damaged or diseased cells in the body. Once cells have differentiated, they have less capacity to form multiple different cell types, and become 'committed' to becoming a particular cell type. Skin stem cells, for example, give rise to new skin cells when needed, to assist regeneration after damage and as part of the normal ageing process.

3. Stem cells are capable of self-renewal. Stem cells are able to divide and produce copies of themselves, which leads to self-renewal. Once a cell has become specialised (has differentiated) to a particular tissue or organ, it has a very limited capacity to self-renew (produce new stem cells) but instead produces only cells relevant to that organ.

July 2010

⇨ The above information is reprinted with kind permission from the Australian Stem Cell Centre. Visit www.stemcellcentre.edu.au for more information on this and other related topics.

© *Australian Stem Cell Centre*

Aborted fetal tissue used in stem cell trial – no thank you

Information from Comment on Reproductive Ethics.

Just knocked down the league table of BBC news today following the announcement of Prince William's wedding plans is the report of a stem cell trial in Glasgow, using fetal stem cells to treat a stroke victim (www.bbc.co.uk/news/health-11763681).

It is always important to get straight to the heart of these stories and not worry if you do not have sufficient scientific knowledge to understand the minute details of what has happened. It's the moral dilemma we have to grapple with.

The claim this time is that by multiplying stem cells from the brains of aborted fetuses and injecting these cells into the brains of patients who have suffered strokes, it may be possible to lessen the effects of the stroke. In the longer term the hope is to perhaps cure these patients.

This particular trial is not in itself aimed at an immediate cure, but is being carried out to assess whether the small dose of fetal brain stem cells initially injected into the patient will be accepted or rejected. The patient will be monitored to see if there are any safety implications and whether there is any indication of improvement.

If the transplant seems safe, other patients will be given progressively higher doses and be monitored over two years to assess results.

A very similar trial is going on in California, but using embryonic rather than fetal stem cells and aimed at patients suffering from spinal injury. The company in question (Geron) is also doing a safety trial to see if the chosen patient develops any reaction to the transplant. Higher dosage of the embryonic stem cells would follow if safety were to be established.

How should we respond to today's news from Glasgow, and the earlier trial report from California?

By going back to the absolute basics of virtuous medicine. Evil may never be done even if good may come of it.

The creation of fetal or embryonic stem cells relies on the destruction of early human life either in the womb or in the test tube. It is irrelevant if permission has been obtained or not. We cannot sacrifice even the tiniest members of the human race, no matter how dramatic or heart-rending the condition of the patients in question. This is a profoundly unethical trade-off.

On the BBC coverage of the Glasgow fetal stem cell trial it is stated that strokes in the UK kill around 67,000 people a year. This is a significant healthcare issue and must be addressed wisely and seriously.

But abortion takes even more lives every year: 200,000 during the pre-birth stages of human development, and it is almost impossible to estimate how many even smaller human lives are discarded during the processes of IVF and stem cell research.

We share the dream of seeking cures for strokes, spinal injury, cancers and all the other conditions which beset modern man, but the realisation of that dream cannot depend on the taking of life of other human beings, no matter how early in development those tiny lives may be.

16 November 2010

⇨ The above information is reprinted with kind permission from Comment on Reproductive Ethics. Visit www.corethics.org for more information.

© *Comment on Reproductive Ethics*

New ethical challenges in stem cell research

Information from the Biotechnology and Biological Science Research Council (BBSRC) and the Scottish Initiative for Biotechnology Education (SIBE).

Deriving sex cells (sperm and eggs) from human embryonic stem cells

Some research groups claim to have made human sperm cells from embryonic or bone marrow stem cells. Many would welcome this, if it gave us a better understanding of what sometimes goes wrong in male fertility. It will not be clear for some time if it could work well enough to enable an infertile man to produce sperm, but what ethical issues would it raise? If a man wanted sperm derived from his own body cells, using stem cells from spare IVF embryos would be no use. He would need a cloned embryo of himself created, say, from some of his skin cells, to make the stem cells and thus the sperm. But should the man produce one or more embryos that are 'twins' of himself, which must then be destroyed to 'create' his own sperm, hopefully to have a child that is biologically his and his partner's (using IVF)? Is there a contradiction if he creates and sacrifices a cloned embryo to create an 'ordinary' one? If iPS cells could be used, that ethical problem would be solved. A second issue is safety. There are risks that the sperm would be damaged during these very complex processes, risking deformed babies. For clinical safety, it would have to be proved, first time and every time, that everything had worked. How could that be done without doing a procedure which treats the child to be born, in effect, as an experiment? This is quite different from a dying patient consenting to take the risk of an experimental therapy with her eyes open. These examples illustrate some of the dilemmas coming from new areas of stem cell research.

Using human embryonic stem cells to test drugs and chemicals for toxicity

Most experts in the field acknowledge that cell replacement therapies based on human embryonic stem cells are a long way away. But another application is coming quickly, which has hardly been discussed publicly and which also raises difficult ethical issues. This is to use human embryonic stem cells (or cells derived from them) to test chemicals for possible toxic effects. Pharmaceutical drugs have to be tested and under new EC REACH regulations so do all existing industrial and household chemicals. Toxicity testing is not on the list of permitted uses of human embryos under the Human Fertilisation and Embryology Act. But having created an embryonic stem cell line, say for diabetes research, can the same cells be used for purposes which would have been refused a licence, like drug screening or even testing cosmetics?

Some say that embryonic stem cells have no moral status. But others disagree because they were created by destroying a human embryo, arguing that the same limits on what the embryo can be used for must apply also to whatever uses the cells would subsequently be put to. It is also possible that scientists would sometimes need to use fresh human embryos to create stem cell lines for drug or chemical testing.

> **Stop and consider:**
>
> Should we allow embryonic stem cells to be used to test the potential toxic effects in humans of pharmaceutical drugs, or cosmetics, or other chemicals?

There are also issues about public accountability and who gets to benefit. When human embryonic stem cell research was first argued for, the big ethical case presented to the public was therapies to treat untreatable degenerative conditions, not for testing chemicals. If this indeed becomes the first major use, has the public been 'kept in the dark'? By law, pharmaceutical companies have to test all intended drugs anyway, but if they could use embryonic stem cells for the rapid screening of chemicals, it would reduce the time and cost of testing. But is a commercial benefit good enough reason for using cells derived from embryos?

Most present toxicity tests rely on using animals. Using human cells instead of animals may give a more reliable test, but would it be better ethically? Both animal research and embryo stem cell research oblige the researcher to look for an alternative. Can we replace one ethically controversial process (using animals) by another ethically controversial process (using cells derived from human embryos)? For example, say you disapproved of using animals to test cosmetics: would it be OK to use cells derived from human embryos instead?

⇨ The above information is an extract from *Stem cells: science and ethics* and is reprinted with kind permission from the Biotechnology and Biological Science Research Council (BBSRC) and the Scottish Initiative for Biotechnology Education (SIBE). Visit www.bbsrc.ac.uk for more information.

© BBSRC

Look, no embryos! The future of ethical stem cells

For years, ethical issues hampered progress in stem cell research. Now, experts believe that developments in reprogrammed 'iPS' cells will truly revolutionise the treatment of life-threatening illnesses.

By Alok Jha

It is unclear at exactly what point the phrase 'stem cell' entered the vernacular, one of very few scientific terms that achieve the status of, say, DNA in not requiring a detailed explanation every time it is written down or spoken.

Whether or not you know exactly what they are or what they do, stem cells imply something very specific: in them is invested the next generation of medicine, revolutionary treatments for everything from Parkinson's to Alzheimer's. On the horizon, there is also the hope of growing genetically matched tissue (even whole organs) to replace anything that has been damaged by disease or accident.

But perhaps the reason stem cells managed to lodge themselves so deep in the public psyche was not just because of their awesome scientific potential, or their ability to turn into the treatments of the future. Perhaps it was politics. For years, stem cells dominated all other science stories in newspaper headlines because they framed an ethical conundrum – to get to the most versatile stem cells meant destroying human embryos.

Research on stem cells became a political football, leading to delays in funding for scientists, particularly in the US. Not that the work itself was straightforward – the process of extracting stem cells from embryos is difficult and there is a very limited supply of material. Inevitable disappointment followed the years of headlines – where were the promised treatments? Was it all over-hyped?

For Paul Fairchild, co-director of the newly founded Oxford Stem Cell Institute, disappointment is just not on the agenda. Over a coffee in the University of Oxford's pathology department, where he is a professor, he explains his vision for the coming, post-hype decade of stem cell science.

'It's an exciting time in stem cell biology for a host of reasons,' he says. 'We've entered a whole new phase in the stem cell field, which has been held up enormously by ethical issues for over a decade.'

Key to this is the discovery, in the past few years, of a way to make stem cells that do not require the destruction of embryos. In one move, these induced pluripotent stem (iPS) cells remove the ethical roadblocks faced by embryonic stem cells and, because they are so much easier to make, give scientists an inexhaustible supply of material, bringing them ever closer to those hoped-for treatments.

> *Research on stem cells became a political football, leading to delays in funding for scientists, particularly in the US*

Fairchild says that iPS technology will 'completely revolutionise the whole of medicine this century'. And the talk is matched by action: the Oxford Stem Cell Institute is a recognition of the importance of the work ahead, a collaboration that brings together 37 laboratories across 17 departments at the university. 'It's an attempt to try and bring all of the work in stem cell biology under one umbrella organisation to allow people to collaborate more effectively.'

Stem cells are the body's master cells, the raw material from which we are built. Unlike normal body cells, they can reproduce an indefinite number of times and, when prodded in the right way, can turn themselves into any type of cell in the body. The most versatile stem cells are those found in the embryo at just a few days old – this ball of a few dozen embryonic stem (ES) cells eventually goes on to form everything that makes up a person.

In 1998, James Thomson at the University of Wisconsin-Madison announced that he had isolated human ES cells in the lab. Finally, these powerful cells were within the grasp of scientists to experiment with, understand and develop into fixes for the things that go wrong.

There were some immediate practical problems with the work. Once scientists had extracted ES cells from an embryo, they could create an immortal line of the cells to use in research. But the wide genetic variations in humans mean that scientists need lots of different lines of ES cells to treat and understand the wide variety of faults. Each new line of ES cells can only be created by fertilising an egg, and these are a precious, rare commodity.

This is also where the ethical problems lie. Extracting ES cells destroys the embryo – for some, this is akin to killing a potential human life. George W. Bush banned the use of federal dollars to support research using human ES cells on all but a limited number of cell lines that already existed in research labs prior to August 2001. Barack Obama reversed that ban, only to be thwarted by a federal court ruling in 2010 that, once again, put the scientists into a state of uncertainty. By contrast, the UK laws are relatively civil – after much debate about the ethics of using human embryos, the Government passed strict, but fair, laws that allowed ES research to go ahead.

But that did not get rid of the practical problems of a lack of donor eggs, or that the experiments with ES cells as treatments – to repair a damaged heart, for instance – were proving unsuccessful.

> **Extracting embryonic stem cells destroys the embryo – for some, this is akin to killing a potential human life**

Enter the iPS cell. In 2007, Shinya Yamanaka at Kyoto University in Japan demonstrated a way of producing ES-like cells without using eggs. He took a skin cell and, using a virus, inserted four specific bits of DNA into the skin cell's nucleus. The skin cell incorporated the genetic material and was regressed into an ES-like cell – it had been 'reprogrammed' using a batch of chemicals in the lab. In a few short experiments, scientists had a near-limitless supply of stem cells that were, seemingly, as good as ES cells for their research.

By creating iPS cells from patients with genetic diseases, scientists have been able to watch which genes go wrong in a variety of conditions, how and when it happens – all of it critical detail in finding ways to stop diseases in their tracks.

The method works like this: take some skin cells from a person with Parkinson's disease and then regress these back into iPS cells. Then coax these stem cells to turn into neurons and watch how they work and, crucially, how they go wrong. These neurons are genetically identical to the patient's own brain cells – allowing scientists to model the disease more accurately and test out ideas or even screen potential drugs.

'It's possible now, of course, to take a few cells from an individual with an intractable disease of some kind for which there are no animal models or any way of studying it, to reprogramme them to an iPS cell state and then differentiate from those the cell type that's affected from the disease,' says Fairchild. 'We now have a way of taking them from that person and studying the cell type itself in vitro that will almost certainly have symptoms of the disease. It's possible to produce these models of human disease in vitro which we've never had access to before.'

Models using iPS cells have proliferated in a few short years: they are now available for, among other things, motor neurone disease, juvenile diabetes and sickle cell anaemia.

'It's exciting to have models that we can now probe and find out what is actually going on for the first time,' says Fairchild. Traditionally, scientists use animal models of disease in order to carry out experiments or test brand new drugs – but these models are not perfect. Using human cells taken from a patient with a human disease is a more promising way of gauging what might happen in people.

Fairchild has been examining how to use iPS cells to modulate the body's immune system. It is a route, he hopes, to less rejection when people get transplants of replacement tissue. The same techniques could even develop into a better way to treat cancer.

He started by differentiating embryonic stem cells into a key part of the body's immune system called dendritic cells. These cells continually sample proteins, called antigens, that sit on the surface of everything in the body. If they determine that the antigen comes from something potentially dangerous or foreign, they set the body's killer T-cells to go after the danger and destroy it. On the flip side, the dendritic cells can also dial down the ferocity of the killer T-cells if they recognise an antigen as non-foreign.

One way Fairchild's team is using this idea is to grow dendritic cells that can recognise the antigens on tumours. Inserting these cells into a patient would dial up the immune system's natural response to that tumour. 'What you do is grow the dendritic cells, then feed them with the tumour-associated antigens in vitro,' he says. 'Then put those back into the patient, so that they're presenting those melanoma antigens.'

> **By creating iPS cells from patients with genetic diseases, scientists have been able to watch which genes go wrong in a variety of conditions**

His experiments have shown that the dendritic cells do indeed stimulate the body's T-cells to respond robustly. The T-cells are then programmed to find the tumour and destroy it. That means a better natural response to cancer, all without the need for toxic chemotherapy.

Another way to use the dendritic cells is to dial down the body's immune response to transplanted tissue or organs. Anyone who has a transplant needs immunosuppressant drugs for the rest of their lives to prevent their immune systems from attacking the new tissue. Stem cells offer the possibility of growing genetically matched tissue for patients, so rejection should, in theory, not be a problem.

But Fairchild points out that growing matched tissue for every individual is likely to be prohibitively expensive for now. Instead, scientists would probably grow replacements from a bank of stem cells that are close matches to the majority of the population. But that means that replacement tissue would still provoke the recipient's immune system.

Fairchild's solution is to grow dendritic cells from the same stem cells being used to grow the replacement tissue for the patient. The dendritic cells can then be used to train the recipient's immune system that the replacement tissue is not to be rejected.

The creativity in the field is exciting, but it is right, also, to point out the many technical issues facing iPS cells in moving outside the lab. Just recently, Andras Nagy of Mount Sinai hospital and Timo Otonkoski at the University of Helsinki found genetic abnormalities when creating iPS cells. In a study published in Nature, they reported the deletion or amplification of certain strands of DNA. 'Our analysis shows that these genetic changes are a result of the reprogramming process itself, which raises the concern that the resultant cell lines are mutant or defective,' said Nagy.

It is just one of a number of research papers raising concerns about the way iPS cells are produced. 'In a laboratory dish, pluripotent stem cells promise to become any sort of cell a researcher could want,' wrote Monya Baker, editor of *Nature Reports Stem Cells*, in a 2009 article looking to the future of the iPS technology. 'But when transplanted into the body, such cells could grow unpredictably: clinicians worry most about tumour formation, although one can imagine other dangers, such as tissues growing in the wrong place.'

The very things that make iPS cells so useful in lab situations – their immortality and versatility – would be disastrous if left unchecked inside people. But, for a field that is barely a few years old, the remarkable pace of achievement should give plenty of hope that these challenges will be met. 'Turning knowledge into medicine is never easy,' says Baker. 'But the pace of knowledge generation itself is fast and furious.'

13 March 2011

© Guardian News and Media Limited 2010

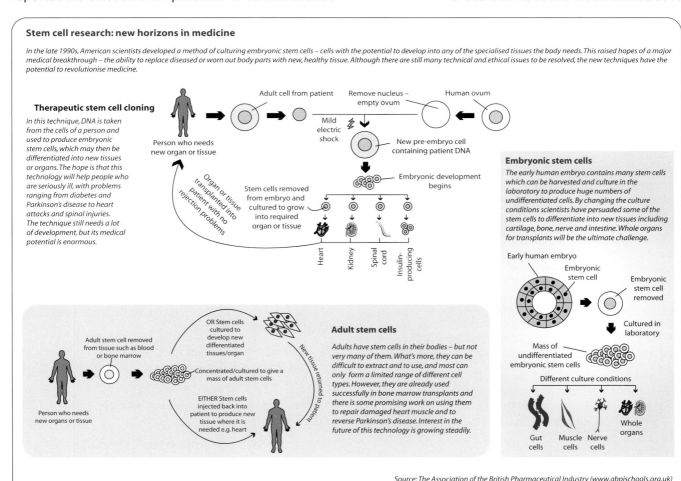

Source: The Association of the British Pharmaceutical Industry (www.abpischools.org.uk)

'I travelled in hope to stem cell clinic'

The XCell Centre holds out hope of a treatment for incurable diseases. Alasdair Palmer, who has MS, travelled to Düsseldorf for a consultation.

Hope is important to everyone, but it is particularly precious to those who suffer from an incurable disease. I am one of that group. I was diagnosed with multiple sclerosis nearly 20 years ago.

I now have to use a wheelchair, and the future is not particularly rosy. So whenever I hear that there has been a new scientific development which could turn out to be a cure, or which could at least halt the progression of the disease, I start to hope.

I know that, on every occasion in the past, the news of a cure has turned out to be false: the claims that the new treatment can reverse MS have, on closer inspection, all turned out to be bogus.

But I can't stop myself from hoping that this time it's different: this time the elixir really has been found.

Like everyone else with a neurological disease, my hopes have been, and are, raised by stem cell research.

In theory, there is every chance that eventually, stem cells – the cells which have the potential to convert themselves into any kind of specialist cell in the body, from brain and nerve cells to the cells that make up your skin or your liver – might eventually provide an effective treatment for MS.

But progress towards that goal is in practice proving to be agonisingly slow, not least because everything that has the potential to do good can also do harm.

It requires a great deal of research to show that a therapy isn't going to be damaging. The history of medicine is littered with treatments that seemed to be miracle cures, but turned out to be harmful.

Even techniques such as heart transplants, which are now routinely practised and save thousands of lives around the world every year, started by killing all the patients who had them.

Nonetheless, when I was asked by the editor of the *Sunday Telegraph* to investigate a medical centre in Düsseldorf that offers stem cell treatment to those who can afford to pay for it, I was immediately intrigued.

I'm aware that a British doctor has been struck off the medical register in this country for offering such treatments. I've also written sceptical articles about the claims made by doctors who say they can provide stem cell 'cures' for MS.

I've also read non-technical summaries of the scientific papers published in medical journals which point to the difficulties with existing stem cell techniques, and to the fact that there is as yet no trial which has provided solid evidence of their benefits.

And yet – there is always the possibility that a brilliant medical researcher has come up with a very effective technique which, while it has not yet been proven to be effective, actually is.

So in spite of my well-rehearsed scepticism, the XCell Centre in Düsseldorf got my hopes up. A quick glance at its website demonstrates that its doctors are very confident that their stem cell treatment is effective in treating MS and other neurological diseases.

My hope focused on a single thought: it is surely not impossible that they have found something that works. And if it is not impossible – mightn't it be worth trying?

I booked myself a consultation with Professor Haberland. I was immediately informed that it would cost 300 euros, and that I would have to pay before the consultation.

The XCell Centre is housed in a large hospital building overlooking the Rhine. As I waited for my appointment, it was clear that everyone else who was waiting for a consultation suffered from the same hope that I did.

Some of them were considerably more desperate for a cure than I am. One woman with advanced Parkinson's Disease – she had travelled all the way from the United States – told me quite simply that she knew the treatment wasn't scientifically proven.

She even realised that there was a significant chance it wouldn't work. 'But what is my alternative?' she asked plaintively. 'For me, there is none, except waiting until I degenerate to the point where I die, or want to die.'

Our conversation was interrupted because I was called in to my meeting with Professor Haberland.

He is a genial man with an air of intense professional competence: he is a trained neurosurgeon who has spent most of his career working on injuries to the spinal cord. He was friendly and patient.

He said that his treatment consisted of taking stem cells from the patient's bone marrow, separating them, and then injecting them.

'We have some amazing results with this treatment,' he told me.

He insisted that the best results came from combining this approach with an operation developed by an Italian doctor named Zamboni, which involved, Professor Haberland said, widening veins in the patient's neck.

Zamboni had, Professor Haberland claimed, established that many neurological conditions were connected to a narrowing of these veins.

When the operation to widen them was used in conjunction with the injection of stem cells, 'you have a win-win situation… 80 per cent of our patients report improvements'.

He pointed to a Dutch patient who had been in a wheelchair prior to the operation, but after it, had been able to walk.

So if I had the treatment, would I be able to walk? 'You would have a chance,' Professor Haberland replied. 'The response is very different from patient to patient. But you would have a chance.' And it would only cost about 19,500 euros.

Against my better judgement, I felt my hopes rising. But there were also some questions. How did Professor Haberland know how successful the treatment was? What was the follow up? 'We ask patients to fill in a questionnaire three months after the operation, and we see them here and test them after six months,' he said.

There is, however, considerable doubt whether a patient questionnaire has any value, since it is not an objective measure.

There is still more doubt as to how many patients actually bother to travel all the way to Düsseldorf for an assessment by the centre six months after the operation.

Moreover, Professor Haberland said he had only been doing the double operation – the combination of stem cells and widening the veins in the patient's neck – for six months.

How, I wondered, could there be any reliable data from such a short period? Would Professor Haberland be publishing his results? No, he said: he had no plans to. 'We need more cases and then we need to make a double blind study in multi centres. This is a long way off.'

But if Professor Haberland hasn't enough patients for a proper trial – how can he be sure even that his operation isn't harmful, let alone that it actually benefits patients?

He was unequivocal that because he uses stem cells from the patient's own body, injecting those cells can't hurt the patient.

It certainly sounds plausible. But no one can know until there has been a large-scale study.

Professor Neil Scolding of Bristol University followed six patients with MS after they had been injected with stem cells taken from their bone marrow for over a year.

Careful monitoring showed that none of the patients were harmed. But the evidence of benefit was less certain: rather than any improvement, it suggested that the patients' condition had stabilised instead of deteriorating further.

It became increasingly obvious in the course of my discussion with Professor Haberland that he does not have any clear, objective and verifiable evidence that his treatment provides the sort of benefits that he claims for it.

It is not even clear that he and the hospital have followed up their patients at all. I asked for any records showing objective evidence of how patients improved after the operation. None have been provided.

That was not enough to extinguish my hopes altogether, however, for Professor Haberland is very persuasive.

He is a subtle salesman. He held out the possibility that I would experience enormous benefits from his operation – but he did not promise them outright. That leaves just enough room for hope to do its work of eroding doubt.

Like everyone else with a neurological disease, my hopes have been, and are, raised by stem cell research

I recorded my interview with Prof Haberland, and showed a transcript to Professor Scolding, an internationally-recognised neurologist and expert on stem cell treatments for MS.

Prof Scolding noted that Professor Haberland's claims about what bone marrow stem cells can do are not, in principle, wrong – but he also pointed out that that is quite different from showing that his treatment can or does make the symptoms of anyone with MS improve.

Professor Scolding was more forthright in his criticism of Professor Haberland's claims for the 'vein clearing' operation developed by Zamboni.

'There are potentially serious consequences for stenting for this so-called condition – one patient has died in the US,' Professor Scolding stated.

'There is emerging published evidence that there is no link between venous drainage in the neck and MS.

'Applying this "treatment" to patients with Parkinson's and other neurological conditions… is extremely eccentric.'

He added: 'If patients had genuinely responded to [Professor Haberland's] evidence-free combination of stem cell therapy and vein clearing treatment, it would be scandalous for him not to have published his results.

'To my mind, if they are persuading often-desperate patients to pay large sums of money for treatments that have no sustainable evidence to support them, it is cynically exploitative and a disgrace to the practice of medicine.'

That, to me, sums it up perfectly. It extinguished whatever hope Professor Haberland had kindled in me.

23 October 2010

© Telegraph Media Group Limited 2010

KEY FACTS

⇨ Biotechnology is the use of biological organisms or enzymes in the synthesis, breakdown or transformation of materials in the service of people. (page 1)

⇨ Some emerging biotechnologies, such as GM crops or synthetic biology, have global implications: they are likely to involve multinational companies and the transfer of money, products and people between countries. (page 4)

⇨ Researchers have managed to transfer human genes that produce useful proteins into sheep and cows, so that they can produce, for instance, the blood clotting agent factor IX to treat haemophilia. (page 7)

⇨ Most of the ethical concerns about cloning relate to the possibility that it might be used to clone humans. (page 8)

⇨ Cloning involves collecting a cell from the animal that is to be cloned (called the 'donor cell') and transferring it into an egg cell that has been removed from another animal. (page 10)

⇨ Nearly half (45%) of the British public say that they would not personally be happy to eat or drink food products from a cloned animal, compared to only 34% who would, a poll has discovered.

⇨ Britain's food safety watchdog has admitted it doesn't know how many cloned embryos have entered Britain after meat produced from cloned cows ended up in food. (page 15)

⇨ While there is no evidence that consuming products from healthy clones, or their offspring, poses a food safety risk, meat and products from clones and their offspring are considered novel foods and would therefore need to be authorised before being placed on the market. (page 17)

⇨ In vitro meat is an animal flesh product that has never been part of a living animal. Cells are taken from a live animal and grown into muscle tissue in a laboratory. These cells form stem cells which are programmed to produce muscle. (page 20)

⇨ The promise of therapeutic cloning is that it will be an effective way to derive embryonic stem cells which can then be used for the development of patient- and disease-specific cell-based therapies as well as the production of stem cells with specific disease characteristics for research purposes. (page 23)

⇨ Human reproductive cloning is the creation of an individual who has identical nuclear genetic material (DNA) to an existing human being, and who is allowed to develop to term and beyond. (page 24)

⇨ The majority of arguments against reproductive cloning have highlighted the possible adverse consequences on individuals, family relationships and society as a whole. (page 26)

⇨ Scientists are trying to find ways to grow stem cells in the laboratory and make them generate specific cell types so they can be used to treat injury or disease. (page 29)

⇨ A stem cell line is a batch of cells that can be grown for long periods of time in the laboratory. These cell lines are grown in incubators under conditions resembling those found in the human body and are commonly used for research experiments. (page 30)

⇨ Some say that embryonic stem cells have no moral status. But others disagree because they were created by destroying a human embryo, arguing that the same limits on what the embryo can be used for must apply also to whatever uses the cells would subsequently be put to. (page 34)

⇨ By creating iPS [induced pluripotent stem] cells from patients with genetic diseases, scientists have been able to watch which genes go wrong in a variety of conditions, how and when it happens – all of it critical detail in finding ways to stop diseases in their tracks. (page 36)

GLOSSARY

Biotechnology

Biotechnology is the use of natural organisms and biological processes to change or manufacture products for human use. Biotechnology is widely used in modern society: for example, in agriculture, pharmaceuticals, the manufacturing industry, food production and forensics.

Clone

An organism which is genetically identical to another. Identical twins are an example of naturally occurring clones, although scientists are also able to create cloned organisms in a laboratory.

DNA

DNA (deoxyribonucleic acid) is the genetic coding which is present in every cell of living organisms. DNA is found in the nucleus of each cell and determines the characteristics for that organism.

Dolly the sheep

Dolly was the first animal to be cloned from an adult mammal – representing a scientific breakthrough. Dolly became very well known amongst the public, but died prematurely due to a disease of the lungs.

Embryo

An animal in the earliest stages of development. In humans, this refers to the eight weeks following conception. After this point the developing baby is referred to as a foetus.

Food Standards Agency

An independent government department set up to protect the public's health and consumer interests in relation to food. The Food Standards Agency is responsible for monitoring the marketing of products from cloned animals and their offspring, in conjunction with the relevant EU legislation.

GM food

GM (or genetically modified) foods are products which have undergone a process of genetic selection to produce a desired characteristic. For example, scientists may transfer the gene for disease resistance from one organism into a genetically unrelated crop, which would result in an improved yield from the modified crop.

In vitro meat

Animal flesh cells that have been grown in a laboratory and which have never been part of a living animal. Scientists are developing in vitro meat as a solution to the health and environmental problems associated with natural animal farming.

Reproductive cloning

The process of creating an organism from the cell of an existing organism, resulting in a genetically identical 'clone'. Scientists transfer the cell nucleus containing the DNA into an unfertilised egg and allow the egg to divide. The fertilised egg is then allowed to develop to full term (in a surrogate carrier's womb in the case of mammals).

Stem cells

There are two sources of stem cells: adult and embryonic. Whereas scientists think that adult stem cells are restricted to maintaining the health of the tissue where they are found, embryonic stem cells have the potential to turn into any cell type. If we can harness their adaptability, they might be a source of healthy tissue to replace that which is diseased or damaged in adults. However, because the embryo is destroyed during the process, the extraction and use of embryonic stem cells is highly controversial. Pro-life groups in particular have been campaigning against stem cell research, as they believe human life begins at conception and embryos should be afforded the same rights as adult humans.

Therapeutic cloning

Also known as somatic cell nuclear transfer to avoid confusion with reproductive cloning, which is a completely different process. Therapeutic cloning involves the creation of embryonic stem cell 'lines' by injecting the nucleus of a somatic cell into an unfertilised egg. Scientists are then able to create many embryonic stem cells, all of which are genetically identical to the DNA from the original nucleus.

aborted fetal stem cells 33
adult stem cells 28, 29
animal cloning 7-21
   for food *see* food from cloned animals
   public attitudes 6
   uses 7-8
animal welfare and cloning 10, 19

biotechnology
   applications 4
   definition 1, 4
   history 2-3

cloning
   see animal cloning; human cloning
consumer reaction to food from cloned animals 11, 12, 14
cows, cloned 12, 13-17
cultured meat 20-21

disease treatment using stem cells 30, 36-7, 38-9
DNA 3
Dolly the sheep 7, 8

egg harvesting for research cloning 30
Egyptians and biotechnology 2
embryo splitting 24
embryonic stem cells 28, 29-30, 35-6
embryos
   moral status 31, 34
   use for stem cell research 31-2
ethics
   and animal cloning 10
   and biotechnology, public attitudes 6
   and human cloning 8-9, 26-7
   and stem cell research 30-31, 34
Europe, attitudes to biotechnology 5-6

fake meat 20-21
fetal stem cells used to treat stroke victims 33
food from cloned animals 10-19
   public attitudes 6
   safety 18, 19

genetically modified (GM) food, public attitudes 5
governance of biotechnology, public attitudes 6

history of biotechnology 2-3
human cloning 22-39
   arguments for and against 26-7
   ethical issues 8-9, 26-7
   legislation 25
   reproductive cloning 22, 24-7
   therapeutic cloning 9, 22-3
Human Fertilisation and Embryology (HFE) Act 25
Human Reproductive Cloning Act 25
hybridisation 2

in-vitro meat 20-21
induced pluripotent stem (iPS) cells 35-7
insulin 3
iPS (induced pluripotent) stem cells 35-7
IVF, use of unused embryos for research 31

laboratory-grown meat 20-21
law
   and animal cloning 11
   and food from cloned animals 12
   and human cloning 25

meat
   from cloned cows 12, 15-17
   laboratory-grown 20-21
Mendel, Gregor 2
milk from cloned cows 12, 13-14
moral status of embryos 31, 34
multiple sclerosis, stem cell treatment 38-9

novel foods 17
Novel Foods Regulation 12

pasteurisation 2
pluripotent stem cells 28, 35-7
potatoes, introduction to Europe 2
psychological consequences of human cloning 27
public attitudes
   to biotechnology, Europe 5-6
   to food from cloned animals 6, 11, 12, 14
   to GM food 5
public ethics and biotechnology 6

regenerative medicine, public attitudes 6
reproductive cloning 22, 24-7
research cloning 32

safety
   food from cloned animals 18, 19
   human cloning 26, 27
saviour siblings 32
SCNT (somatic cell nuclear transfer) 24
seed hybridisation 2
social consequences of human cloning 27
somatic cell nuclear transfer (SCNT) 24
stem cell lines 30
stem cell research 29-37
   ethical issues 30-31, 34
stem cell therapy 30
   multiple sclerosis 38-9
   stroke victims 33
stem cells 28, 29-30
   embryonic 28, 29-30, 35-6
   from aborted fetuses 33
   iPS 35-7
   tissue specific (adult) 28, 29

stroke victims, treatment using fetal stem cells 33
synthetic biology, public attitudes 5

therapeutic cloning 9, 22–3
    research applications 23

tissue specific (adult) stem cells 28, 29

UK, animal cloning 10
US, animal cloning 10–11

Wallace, Henry 2

ACKNOWLEDGEMENTS

The publisher is grateful for permission to reproduce the following material.

While every care has been taken to trace and acknowledge copyright, the publisher tenders its apology for any accidental infringement or where copyright has proved untraceable. The publisher would be pleased to come to a suitable arrangement in any such case with the rightful owner.

Chapter One: Biotechnology

What is biotechnology?, © Association of the British Pharmaceutical Industry, *Emerging biotechnologies*, © Nuffield Council on Bioethics, *Europeans and biotechnology in 2010: winds of change?*, © European Union, 1995-2011.

Chapter Two: Animal Cloning

Animal cloning, © Understanding Animal Research, *Adult cell or reproductive cloning* [diagram], © The Association of the British Pharmaceutical Industry, *Timeline of domestic species cloned*, © Compassion in World Farming, *Q&A on cloning of animals for food*, © Compassion in World Farming, *Cloned cows*, © YouGov, *'If anything, this milk will be better quality'*, © spiked, *Food watchdog admits tracking cloned cows 'impossible'*, © Telegraph Media Group Limited 2010, *Cloned animals and their offspring*, © Crown copyright is reproduced with the permission of Her Majesty's Stationery Office, *Cloned British meat is 'safe'*, © Telegraph Media Group Limited 2010, *More evidence required on cloning*, © Soil Association, *In vitro meat*, © Compassion in World Farming, *Why I'd happily eat lab-grown meat*, © Guardian News and Media Limited.

Chapter Three: Human Cloning

Therapeutic cloning (somatic cell nuclear transfer), © Australian Stem Cell Centre, *Reproductive cloning*, © BioCentre, *What are stem cells?*, © Australian Stem Cell Centre, *Stem cell research: hope or hype?*, © Irish Council for Bioethics, *First 'saviour sibling' stem cell transplant performed in UK*, © BioNews, *Aborted fetal tissue used in stem cell trial – no thank you*, © Comment on Reproductive Ethics, *New ethical challenges in stem cell research*, © Biotechnology and Biological Science Research Council, *Look, no embryos! The future of ethical stem cells*, © Guardian News and Media Limited, *Stem cell research: new horizons in medicine* [diagram], © The Association of the British Pharmaceutical Industry, *'I travelled in hope to stem cell clinic'*, © Telegraph News and Media Group Limited 2010.

Illustrations

Pages 1, 15, 21, 33: Angelo Madrid; pages 2, 13, 20, 26: Simon Kneebone; pages 4, 9, 19, 25: Don Hatcher; pages 6, 17: Bev Aisbett.

Cover photography

Left: © clix. Centre: © Andrew Petrie. Right: © Daniel Shepherd.

Additional acknowledgements

Editorial by Carolyn Kirby on behalf of Independence.

And with thanks to the Independence team: Mary Chapman, Sandra Dennis and Jan Sunderland.

Lisa Firth
Cambridge
April, 2011

ASSIGNMENTS

The following tasks aim to help you think through the issues surrounding the cloning debate and provide a better understanding of the topic.

1 Brainstorm to find out what you know about cloning. Why are scientists interested in cloning? Why is it controversial?

2 There are many science fiction films which deal with the subject of cloning: for example, 'The 6th Day', 'The Island', 'Blade Runner' and 'Moon'. Why do you think this is such a popular theme? Watch one of these films and write a review.

3 Carry out a survey to find out whether the pupils in your class would eat meat from a cloned cow. Write an analysis of your findings, including graphs to display your results.

4 The laws which regulate cloning are constantly being reviewed and updated. Visit the Food Standards Agency's website to find out about current legislation and advice. Do you think the Food Standards Agency are successfully regulating animal cloning for food?

5 Read *In vitro meat* and *Why I'd happily eat lab-grown meat* on pages 20-21. Imagine you are a part of a research team which is trying to introduce in vitro meat into supermarkets. In groups of four, create a TV advert which informs viewers of the benefits of lab-grown meat compared to farmed animal meat. You will need to reassure consumers of the safety of the product and make it sound appealing. What sort of stigma do you think would surround the idea of in vitro meat? How could you encourage consumers to try the product?

6 Read *Stem cell research: hope or hype?* on page 29 and look at the diagram on page 37. In pairs, create a PowerPoint presentation that explains what stem cell research is and what it is used for. Try to condense the information into six to eight accessible slides.

7 Study the graph on page 23. What do these statistics show? Which of the three research areas are the most/least popular? Why do you think this might be?

8 In groups, create two spider diagrams on A2 paper. One should include all the reasons you can think of to support the idea of developing human cloning research, whilst the other should present the arguments against human cloning. Which diagram is more convincing?

9 Read *Emerging technologies* on page 2. Which do you think are the ten most important pieces of information in the article? Condense these ten points into concise bulleted list.

10 Read *Stem cell research: hope or hype?* on page 29. What is your opinion of stem cell research? Do you think it is ethically acceptable to use embryonic stem cells for medical research? Now read *I travelled in hope to a stem cell clinic* on page 38. This gives a personal account of someone who hopes to benefit from stem cell research. Has your opinion changed since reading the second article? Discuss your answer in groups.

11 Read Jodi Picoult's novel 'My Sister's Keeper'. How does the novel explore the issues of 'saviour siblings'? Who do you sympathise with more, Anna or Kate? Do you think the novel leads the reader to a moral judgement of the characters or leaves it open for you to draw your own conclusions?

12 In groups, role play a radio interview on the topic of animal cloning for food. One member of the group should play the role of the host, one a scientist keen to pursue animal cloning as a method of food production, one a representative of a faith group who opposes cloning, one a member of a lobby group which addresses the world hunger crisis, one a representative of an animal welfare group against the idea of animal cloning and one a member of the public who is yet to be swayed either way. Debate the topic for around 15 minutes.

13 Write an article for a magazine which explains the differences between therapeutic cloning and reproductive cloning in layman's terms. Members of the public often assume that the term 'cloning' refers solely to the genetic replication of organisms. Your article should dispel the myths surrounding the topic of cloning and inform readers of the processes involved with therapeutic cloning and its uses, compared with reproductive cloning.